Science International is the first history of a worldwide organisation of scientists, now involving thousands of participants, which was started a century ago when a few visionaries founded the International Association of Academies (1899–1919). This was succeeded by an International Research Council (1919–1931), which, in 1931, became the International Council of Scientific Unions (ICSU). The initiative to have an international arena for scientists survived two global wars, as well as immense political, economic, and social change in the 20th century. This history describes how national scientific academies as well as International Unions of scientists from specific disciplines learned to work together. From these alliances sprang great co-operative projects such as the International Geophysical Year and the International Biological Programme. Today ICSU is a global scientific organisation directed to the study of the entire planet and prospects for the human race. This detailed account will appeal to researchers in the history of science who are interested in the organisational aspects of science in the 20th century, and to professional scientists working in the service of science on a national or international level.

A black and white reproduction of a water colour painting of the Hôtel de Noailles, ICSU's International Secretariat in Paris.
By Professor Keiji Higuchi of Japan

Science International

Top The beginning of the International Association of Academies. An international gathering of European academic men of science. Wiesbaden 9 October, 1899. (back) Lieben, Schuster, Dyck, Armstrong, Leo (middle) von Sicherer, Windisch, Rücker, Bowditch, Wislicenus, Famitzin, Moissan, von Lang, Ehlers (front) Darboux, Auwers, Newcomb, von Zittel, Gomperz, Mussafra

Bottom A mature ICSU: a 1985 gathering at Schloss Ringberg: pure and applied science, industry, all the continents, men and women.

Science International

A history of the International Council of Scientific Unions

FRANK GREENAWAY

CAMBRIDGE
UNIVERSITY PRESS

CAMBRIDGE UNIVERSITY PRESS
Cambridge, New York, Melbourne, Madrid, Cape Town, Singapore, São Paulo

Cambridge University Press
The Edinburgh Building, Cambridge CB2 2RU, UK

Published in the United States of America by Cambridge University Press, New York

www.cambridge.org
Information on this title: www.cambridge.org/9780521580151

First published 1996
This digitally printed first paperback version 2006

A catalogue record for this publication is available from the British Library

Library of Congress Cataloguing in Publication data
Greenaway, Frank.
   Science international: a history of the International Council of
Scientific Unions / Frank Greenaway.
       p.   cm.
   Includes bibliographical references and index.
   ISBN 0 521 58015 3 (hc)
   1. International Council of Scientific Unions – History.
I. Title.
Q10.I69G74   1996
560'.01–dc20   96-31561
                                                          CIP

ISBN-13 978-0-521-58015-1 hardback
ISBN-10 0-521-58015-3 hardback

ISBN-13 978-0-521-02810-3 paperback
ISBN-10 0-521-02810-8 paperback

Contents

Contents

Foreword

In 1990 the Executive Board of the International Council of Scientific Unions decided that the 60th anniversary of the inception of ICSU in its present form should be marked by the publication of a history. It was clear that a substantial work could not be prepared in time for publication in 1991, so Lars Ernster, a former Secretary-General, was invited to prepare a commemorative issue of the ICSU magazine *Science International* which appeared in September 1991. Frank Greenaway, a former officer of the Science Museum London and former Secretary-General of the IUHPS, was then invited to take on the major task of writing a book.

The Board believes that this first attempt at writing a history of ICSU is not only a valuable contribution to the history of science, but also an important instrument for explaining what ICSU is and does to those working in ICSU and to those working with ICSU.

The Executive Board offered all possible help, but wished that the final text should be the work of one hand, a personal view. ICSU is highly indebted to Frank Greenaway for the way he interpreted and executed his task, always seeking to maintain high standards with respect to historical accuracy and readability. Thanks are further due to Maurits la Rivière who was asked by the Board to oversee the project, and to L J Cohen, J-C Pecker, and G F White, whom the Board asked to read the final draft and to comment on it.

(J C I Dooge, President)

Preface

Our subject is really the betterment of the human condition, but our material is individual scientists choosing to work together and trying to take science beyond the limitations of individual effort. The number of scientists in the world, acting in many ways, increased greatly in the 20th century. Of them an influential few (relatively few but amounting to some thousands) took part in the events set out here which have involved members of a distinctive type of international organisation. Two words occur repeatedly in these pages: *Academy* and *Union*. Each has been used many times to label a voluntary gathering of men and women of like interests. The Academies discussed here do not always bear that word in their titles, but all represent the scientific communities of individual nations, men and women of several callings working together to give the people of the national communities to which each owes his or her way of life a single scientific voice. Groups of scientists all over the world, of a single disciplinary commitment, have devised ways of working together regardless of their personal national affiliation, the international communities they have developed using the title of Union. Such Unions can exist independently, but what we are concerned with here is the way the Unions, through the International Council of Scientific Unions, have developed a way of working with each other and with the national academies. Most of the increasing number of personal participants have been occupants of academic posts, in universities and research institutes, but they have been men and women whose interests lay well beyond the individual pursuit of abstract knowledge. The base of their involvement has been that they have cared for the effective sharing of new knowledge, within a particular discipline and, more important, between disciplines. The peak of their involvement is their concern, not always expressed or felt, for the common future of humankind, which has meant that they have created lines of communication, not only with each other, but also with the world of politics and economics. This book has to begin long before the present form of ICSU and must deal with its predecessor bodies, the International Association of Academies and the International Research Council. Indeed, as the story begins some of the nations and their scientific academies did not exist. We can therefore only go as far as the present position and hold back from speculation. So much is seen to have changed that it would be unwise to predict what might happen next.

Acknowledgements

As with any work of this sort, many Librarians and Archivists must be thanked, notably those of UNESCO, the Royal Society of London (Mrs Sheila Edwards and her colleagues, particularly Sandra Cumming and Mary Sampson), the Science Museum Library (South Kensington), the Royal Institution of Great Britain. Members and staff of many academies have contributed, notably the National Academy of Sciences of the USA, the Académie Royale de Belgique, Royal Netherlands Academy of Arts and Sciences, the Royal Swedish Academy of Science, the Austrian Academy of Sciences.

Viviane Quirke (Oxford University) acted as Research Assistant for some sections. Parts of the text were read by Brigitte Schroeder-Gudehus (University of Montreal), William Shea (McGill University), Peter Sabine (formerly Deputy Director, British Geological Survey, who commented on geological matters, R J W Keay (formerly Executive Secretary, the Royal Society of London) who commented on biological and administrative matters. There were many conversations and much correspondence with, among others, D G Chisman, Phyllis Glaeser, Roger Hahn, Philip Hemily, Eric Kupferberg, Nicholas Kurti, Sir James Lighthill, J J McCarthy, Henry Metzger, Sir Frederick Warner, Peter Warren, E Barton Worthington, John Ziman.

From beginning to end, Professor J W M la Rivière (formerly Secretary-General of ICSU) watched and sustained all the work on the book. Many of his suggestions, often substantial, found their way into the text, as did many other suggestions by the learned professional staff of Cambridge University Press.

Above all, thanks are due to the permanent staff, past and present, of ICSU itself, who have never failed in their support over text content and the mechanics of getting words onto paper. It was useful at many points to be in touch with F W G Baker, former Executive Secretary of ICSU. In the early stages, M L Millward took a lead, which on his leaving ICSU, passed to Mrs P ('Tish') Bahmani Fard, who then kept in constant touch with all aspects of the work. Throughout, no author could have been better supported than by Julia Marton-Lefèvre, Executive Director of ICSU, to work with whom, and her colleagues, has been a privilege.

Chapter 1

International Association of Academies
(1899–1914)

Science has always been, to an increasing extent, international in its need for shared thought and for joint action. The importance of its international character has steadily increased. We cannot fully understand the way science advances nor the way in which it affects our lives without examining that international factor in its make-up. Equally, science is affected by events outside the world of science, which is not a closed world but one which interacts with individuals and with the communities they create.

Until the middle of the 19th century, the international dimension was not yet as strong a contributor to the development of the large mass of scientific knowledge, nor as vigorous a stimulus to co-operative action, as it has become in the present century. It was for a long time more a question of communication, and mutual enlightenment. However, from the middle of the 19th century useful modes of collaboration become clear, in fields related to our global existence: such as astronomy, geology, geodesy, cartography, and the several aspects of biology. The same is true of factors in our personal lives, as for example in the settlement of standards and measures of time. Units of length, weight and volume needed to be agreed quite early, and, as society became more integrated, the need for precision and for wider acceptance grew. Then, as the discoveries in electricity of the early 19th century began to be applied and were seen to be widely useful, totally new kinds of measures and their units had to be devised and agreed. Even in their earliest primitive forms these matters of measure have always been of common concern to both thoughtful and practical men. The development of their scientific treatment and the desire to work on them grew together. The internationalisation of biological nomenclature had already been well founded in the 18th century.

It all looked simple and desirable. In 1918 a report of a committee of the Royal Society, making plans for a meeting to consider a new Association, said of associations:

> Some are intended to establish uniformity in the standards of measurements. Some are intended to advance science by co-operation. Some merely encourage personal interchange of opinion. Of those which aim at directly organising scientific progress, some do so because it is considered essential to co-ordinate observations taken in different places, while some aim just at economy of labour.[1]

This statement would serve well enough as an introductory guide to the philosophy of the international associations we are looking at here, but, while such an analysis could

1

reflect the experience of the 20 or so years from the last decade of the 19th century, it is far from sufficient to describe the complexity of the international system of the last decade of the 20th century.

In the 20th century the scientific community created a world-wide system of mutual co-operation in discovery and in application. Trade, industry, and transport were involved, as were new methods of publication, traditional and novel, of transport and of electrical and electronic communication. There was political recognition of science as a factor in economic and cultural life, so governments come into play.

Conversely, science has turned to government funding where private resources fail. Some scientific activities are generated within the machinery of government, not only for the obvious military purposes, but also for welfare or taxation.[2] It is possible and legitimate to view this history in sociological terms.[3] Indeed, it is unwise to neglect this dimension since science does not exist in a sociological vacuum. However, some recent writers have recounted, and commented on, the history of the early international scientific organisations with little regard to its scientific components. In such a study as this one must not let 'scientific' sink out of sight in 'international'.

The most widely active and influential body on this world-wide stage of scientific co-operation is the International Council of Scientific Unions. The history of ICSU is a part of world history. It is presented here as a story full of organisations, of discoveries and inventions, of co-operation and controversy. But, since, as a wise man said, 'History is the essence of innumerable biographies', it is also full of people.

Science classical

This is not a tale where one can say 'To begin at the beginning.' The beginning is too far back, as far back as you like to go, into classical antiquity, into the rise and fall of Egypt, or of the succession of Chinese cultures,[4] into the rise of Islam and its links with the Western Europe of the Middle Ages or of the ancient speculations and achievements of India.[5] By the middle of the 19th century science was a world phenomenon, a heritage possessed in pieces by many groups, some individual, some co-operating groups, some linked with states and nations. Although the need to communicate had always been acted upon, it became apparent to some in the later part of the 19th century that it was time to make some co-ordinated effort to unite the entire scientific heritage into a shared whole. So let us make that our beginning, around the time of the great discoveries in such areas as mathematical physics, in industrial chemistry, in the physiological basis of heredity which occurred towards the close of the 19th century.

It was partly as coincidence, partly as consequence, that we see efforts to develop internationalism in science at a time when the practitioners of several sciences had become conscious of new prospects. The developments of mathematical physics at the hands of Helmholtz, Boltzmann, Kelvin, Maxwell, and others were matched by the discoveries in radiation of such men as Röntgen and Hertz, in matter theory of J J Thomson and Becquerel. In chemistry, the consolidation of an atomic theory revealed elegant order in universal matter, and the powerful methods of organic synthesis greatly extended the variety of beneficial substances. In biology, new ideas had grown out of the acceptance of Darwin-based theory, while, at the other end of the scale of living matter, new microscopic techniques and new biochemical equipment enlarged one's vision of the features underlying the life process itself.

In applied science, such advances as the understanding of the nature of metals or the application of advanced mathematics to engineering were not only altering the comforts of life in the advanced countries, but also promising new life the world over. At the same time, applied science was exhibiting its power to threaten. The complexity of the moral dilemma is illustrated by the life of Alfred Nobel and his legacy.

It was in the area of communications that science had most strongly conferred benefits on itself. Scientists have always communicated, by whatever means were available. In the era of Kepler and Tycho Brahe, personal contact was essential. Although much science developed as separate disciplines, much progress was necessarily made when discussions were interdisciplinary, as for instance in astronomical or geodetic matters. The study of the correspondence of such dedicated communicators as Oldenburg, Mersenne, and Leeuwenhoek in the 17th century, or Benjamin Franklin in the 18th, occupies many a happy scholar for years on end. The travels of scientific men occupy large parts of their many biographies, with personal contacts sometimes assuming great significance, as for example in the encounters between Davy and Gay-Lussac.[6] In earlier times movement was slow but unimpeded, and the present-day scientist may well be astonished to learn that Davy could travel to Paris at a time when Britain and France were at war. In some fields, notably astronomy (e.g. the transit of Venus in the years 1761–9) and geodesy (e.g Biot and Arago's work on Spanish territory for measurement of an arc of the Paris Meridian in 1806), international collaboration was essential because related observations had to be made over distances that went across borders or into distant territories.

The conditions for scientific co-operation are thus created in part by science itself. In the 19th century the means of communication changed radically. The telegraph, then the telephone, allied with the railways, made it easy to send messages, to speak tongue to ear, and to meet face to face, to be complemented by air transport and electronic communication in the 20th century. We are concerned here with the world of science; but not only must we deal with the role of learning at large in this first chapter, we must not lose sight of it at any time.

The Academies

The word *Academy* (*Académie, Akademie*) came into modern languages from Plato's gathering of friends in a garden to discuss common philosophical interests. It was a useful label for the kind of gathering which groups of people with enquiring minds began to form around many centres from the Middle Ages onwards. Most were local in their affiliation, and such academies of limited territorial commitment grew up in many of the small kingdoms and principalities which characterised European government for centuries. With the rise of some larger nations, however, there developed national academies (not always bearing that word in their titles). Academies of local commitment and academies of national authority both played a part in the events of the late 19th century we come to later.

The creation of national academies of science is spread over a long period. By the middle of the 19th century there had grown up many national organisations for scientific collaboration, some venerable national academies like the Royal Society of London and the French Académie des Sciences, rooted in the 17th century, being taken as models in many countries. The dating of 'foundation' of many academies is often imprecise because in most cases the formalities were preceded by informal gatherings, the success of which confirmed the view of some group of enthusiasts that their colloquies could be systematised. The Royal Society of London was the earliest (1660) to have taken on a form and status capable of surviving without interruption into modern times.[7] The Académie des Sciences followed it in 1666 to undergo, during the political and constitutional upheavals of the end of the 18th century, changes which never diminished its central role in French science.[8] In Italy the Accademia del Cimento and the Accademia dei Lincei (those whose sight was keen as that of the lynx) had risen and been dissolved, the Lincei to be revived again in the 19th century and continue into modern times. In Belgium one sees the kind of political and sectarian complications which are to be encountered in many countries: the formation of an Académie Royale des Sciences et des Beaux Arts de Belgique, later divided into separate French language and Flemish language entities. We see the formation of academies of science by royal decree in Russia in 1725, in Sweden in 1741, in Denmark in 1742, in Spain in 1847.

The German situation is unique, and was to influence events in the international pattern of science at several periods in the late 19th century and in the 20th. In Austria an Akademie der Wissenschaften was founded in Vienna in 1846. In the German territories, consisting of a large number of autocratic principalities, many academies of greater or less importance were created, mostly reflecting the personal interests of rulers. Some were important: the Bayerische Akademie der Wissenschaften dates from 1759, the Preussische Akademie der Wissenschaften was formed first in 1700 and revived in a more permanent form in 1812. There were some academies associated with particular towns, such as Göttingen, a dependency of Prussia which tolerated a degree of independence, and Leipzig, principal town of Saxony which, like Bavaria, retained a nominal independence under the German Empire, into the early 20th century.[9] No two of these German academies were alike in constitution or concerns. Their desire to strengthen their own intellectual capacity plays a very important part in the history of the origins of the predecessors of ICSU.

In other parts of Europe, academies were as scattered as the national remnants of the consequences of the Congress of Vienna. We are not concerned here with earlier 19th-century international politics except when it is needed to identify divisions in the world of learning. One later political change needs comment: the opening up of Japan following the Meiji Restoration of 1868. The changes in Japan were rapid and militarily astonishing. In the wider field of academic science its day was yet to come, but in one field it made a virtue of misfortune: its many earthquakes. An Englishman (J J Milne) was able to stimulate the interest of Japanese scientists in work they were to encourage and develop.[10] This led to another strand in international scientific co-operation, namely in seismology which finds its phenomena in many parts of the world. The Japan Academy was founded in 1879.

Another more recent creation in a leading nation was that of the United States National Academy of Sciences (1863). The distinguished American Philosophical Society in Philadelphia had already been in existence as a private body for a century, and its influence was very great. Thomas Jefferson, third President of the United States seems to have been prouder of his Presidency of that body than of the Presidency of his nation.[11] But the United States government, becoming conscious of its role on the world scene, decided that the lead, at least nominally, in American science should be in the hands of a national body, so in 1863 the National Academy of Sciences was founded.[12]

The earlier societies had embraced men of science of all interests, modern distinctions between personal commitments and the compartmentalisation of science not yet being a feature of the scientific scene. Even the terminology of science had not taken on its modern form. The English word *scientist* was itself popularised by the Englishman William Whewell, replacing 'natural philosopher'as a name for someone interested in reflection and research on the condition of the material world. Even now it does not have an exact equivalent in French or German.[13] But specialism in subject-matter and in commitment grew rapidly in the first half of the 19th century, so that national societies with specific fields of interest began to be formed, like the several Chemical Societies.[14] It even began to be seen that there was bound to be specialisation within such distinct areas of science[15] and that a threatened fragmentation of science needed to be countered by a contrary influence: a mutual appreciation engendered by organisations whose object would be the gathering together of men united by common feeling rather than by common technique.

The expanding literature of science[16] and an expanding scientific instrument industry[17] provided for a large degree of sharing of techniques, but the habit of talking together was irrepressible and, indeed, irreplaceable. An example of the union of theory and experimental technique is seen in the Loan Collection of Scientific Instruments of 1876, a unique and original assemblage of a large variety of current instruments held in London. Associated with it was a series of 'Science Conferences': lectures by devisers and users of many of the instruments exhibited, in which the interpretation of results was made all the more profound and intelligible by each speaker being able to draw attention immediately to the means by which he had attained them.

Some societies did grow out of concentration on a limited subject interest. One of the earliest was the Magnetische Verein[18] set up by Gauss[19] and Weber.[20] A great many magnetic observatories already existed, but the work of these two men brought them together rather closely, so that there could be extensive publication. This body had a limited life mainly because of personal and political difficulties, but it stands as an early example of an attempt at a global observation system. It was concerned, that is to say, not only with exchange of information, but also with co-operation in observation, and the merging of individual sets of observations. These two factors are central to the growth of international scientific bodies from that time onwards.

Organisation and organisations

Most academies were concerned with bringing scholars together, regardless of their main personal interests. Other kinds of organisation grew up which brought together specialists in particular sciences. Sometimes the subject demanded international collaboration.

A prime example of such a fundamental issue is to be found in geodesy, a science of which the root concern is the figure of the earth, globally and locally. It had presented questions since ancient times, and had become a distinct study from the time of Columbus and the first great circumnavigations. Advanced mathematics was applied to observations in the 18th century, and became an essential part of the science in the 19th (e.g. Bessel functions).[21] In 1862, following a Prussian Army initiative, some of the German states developed a Central European Geodetic Association (Mitteleuro-päische Gradmessung) at the first general conference of which 13 countries were represented. Many were German states, reflecting the academic autonomy they still enjoyed. The growing European character is reflected in the change of title in 1867 to European Geodetic Association. In 1883 a world character was assumed with invitations to Great Britain and the United States of America, and discussion of a common prime meridian and an international time system. In 1886 the name was changed again to International Geodetic Association. (In many such international bodies one often observes the strong influence of some one institution or person, in this case F R Helmert[22] of the Prussian Geodetic Institute). The last General Assembly was in Hamburg in 1912. The geodesists continued to play a prominent part in international science.

Bit by bit there developed the habit of scientific conferences, some being regular meetings of groups to review progress in some particular field, some specifically to examine an outstanding problem, such as the Karlsruhe Chemical Conference of 1860, which had as its aftermath the clarification of the system of atomic weights, a development which had a profound influence on the whole of chemical science.[23] The botanists were already holding International Congresses.

There is another side to this: one could multiply examples of the birth of organisations. No two would be alike. Consider two astronomical examples. The first International Astronomical Congress was held in Heidelberg in 1863. In 1887 the first International Conference on Astronomical Photography was held in Paris, marking the way in which advance in experimental or observational technique can so change the profile of a field of study that new institutions are needed to support the new studies they generate.

We can see another example of the concentrating effect of novel techniques in the work of Etienne-Jules Marey, who became devoted to quite new types of study of physiology and most notably to the use of repetitive photographic techniques for the study of animal movement. In 1898 he succeeded in obtaining support, not only from French sources but also from the Royal Society and other foreign academies, for a new physiological institute eventually named after him[24]

The common observational and computational services had more than an abstract scientific significance. Navigation and cartography had been major practical problems since the 15th century. The international work in geodesy led to collaboration in the determination of latitude and longitude and later, through the Bureau International de l'Heure, in setting standards of time. Many of the bodies created were related by intergovernmental diplomatic conventions, which were adequate so long as good relations existed. Many of these earlier conferences and international contacts were sporadic and individualist, although there were some notable long-lasting series, such as the Geographical Congresses.[25]

There were also some more specific endeavours. One was the inspiration of a man who died before his vision came to fruition: Karl Leyprecht. He conceived the idea of an International Polar Year, in which observers from many countries would work together, all exercising their own special skills and resources, but pooling results to reveal more than could be found from any one study on its own. Leyprecht never saw the International Polar Year of 1882–3, but in many senses it was his year, and the changes that were to come about owed much to his initiatives, and the demonstration that collaboration could be achieved on a grand scale. His vision had not faded 70 years later (see Chap 12 on IGY).

Then there began, in the last decade of the century, a movement to establish some wider unity of action. It becomes apparent first between a few of the leading German academic institutions. There were many German bodies, some set up by governments, some associated with universities, but they found it easy to seek common ends, partly because of a common language, but also because of recognisably common cultural roots. They gave adequate individual support to some kinds of scientific work, but the need for collaboration in the common needs of scholarly study stimulated a few of the most distinguished academies to create an organisation with limited but important aims, a 'Cartel'. The Cartel concept was very influential in the German states at this time, playing an important part in the development of industry and trade. Just like the members of an industrial Cartel, the members of this new, very limited, academic organisation looked inwards to its German community needs. However, they also looked out, and became conscious of the limitations of their perspective.

Membership of these German academies reflected German history and the persistence of local commitments. You could be a full member of an academy only if you resided in its home territory. The Cartel had its first meeting in 1893. Its membership was narrow: the Academies of Göttingen, Leipzig, Vienna and Munich. The Berlin Academy (Königlich Preussische Akademie der Wissenschaften) remained aloof. There were, however, guests from abroad, including Fellows of the Royal Society.

Conversations began, in the corridors, about extending the Cartel, in fact or in concept. The detail of what was said did not find its way into the archives, but over the next few years the idea began to grow and to change.

The archives and formal reports of the Royal Society, as well as later recollections[26] show that there was a good deal of correspondence and personal contact. Among the names that appear are those of Wilhelm His[27] who later wrote an account of these events. In a lecture[28] given at the Royal Institution in 1906 Schuster referred to the influence of the aged Mommsen,[29] the grand old man of German humanities, a reminder in this present history of a scientific organisation that at the turn of the century scholarship was not yet quite as divided as it seems to be now. Be that as it may, there was needed some new initiative if something more comprehensive than the Cartel were to be created.

An invitation to progress

If we are put a date to the inception of the modern pattern of international collaboration and identify one year we might perhaps put it at 1897, and take note of an invitation offered by Eduard Suess.[30] He was President of the Austrian Academy of Sciences, a notable geologist. He had been developing with colleagues the extension of the idea of a Cartel, nothing less than the creation of an Association of Academies. This association would draw academies of other nations into closer connection with the Cartel Academies (Göttingen, Leipzig, Vienna, and Munich). The Berlin Academy remained aloof. Although he does not later play a large part in this story, we should remark that Suess was typical of the kind of man we find taking an interest in international developments: academically distinguished, widely travelled, possessing an exceptional command of languages (he had been born in London and learned his first English there, where his father had for a time been in business), of wide political and administrative experience. Men like him turn up again and again. On behalf of his colleagues he wrote to the Secretary of the Royal Society inviting it to send representatives to attend a meeting of representatives of the Cartel in Leipzig in 1897. The invitation was quite formal. Although Schuster later described his attendance as private, perhaps meaning that they were there as observers, not yet considering any Royal Society involvement, it was not long before that involvement developed.

Other correspondence followed, a larger Royal Society representation (Michael Foster, Rücker, Armstrong, Schuster) attended a further meeting in Göttingen, where, no doubt, the beneficent influence of Gauss was still felt. This was still only a Cartel meeting. However it did begin a sequence of international activity which, although interrupted by two World Wars, and transformed in many ways, continues to the present day.

It must be emphasised that the intention was not at first international in the present-day sense.[31] Suess's 1897 invitation to the Royal Society was to that body alone. The Royal Society's response, to the 1898 invitation was positive but raised an important matter. The delegates were instructed how to react if any question of the adherence of the Royal Society to the Association of Academies and Scientific Societies was brought up. They were to make it clear 'that such adhesion must be contingent on the Association being rendered truly international by the adhesion of other nations, and more particularly of France'.[32] Such a requirement enlarged the vision of the proposal far beyond the original Cartel concept. The Cartellists may have seen this international view as inevitable, but they had a German problem. The Franco-Prussian War still cast a shadow over relations between the French and the German communities. A direct approach from the Cartel to Paris was hard to contemplate. The Royal Society had no such inhibitions and could speak freely.

The meeting gave much time to productive discussion of the kind of question which had already become characteristic of international co-operation, the global issues of geodesy, the professional value of co-ordinated information (an International Catalogue of Scientific Literature),[33] pooling of information on nationally organised exploratory expeditions, as well as topics in the fields of literature, history, and archaeology . The most important question, however, was the limited objective of the extension of the Cartel. There seems to have been a deep feeling (that turns up again and again) that personal contact was the most important contribution any scientific organisation could make to the progress of science.

The terms of the resolutions which concluded the business foreshadowed issues which were to be debated ever after. Their scientific content was small compared with that of those which were to take place in the successor bodies, but these administrative matters were to go on playing their part in the evolution of the later bodies, the International Research Council and ICSU itself.

The Cartel (that is, the inviting body) resolved:

> The Delegates of the Cartel will propose to their respective Academies to accept in principle the institution of an international association of important societies, and to empower the present local authority (Vorort) of the Cartel to communicate the result to the Royal Society.

> The Delegates of the Royal Society are requested to submit to the Royal Society a similar proposal, and to communicate the decision of the Royal Society to the present local authority

It was further agreed that, should the principles of these resolutions be accepted, the best method of proceeding would be for the Royal Society to ascertain the views of the Académie des Sciences at Paris, of the Imperial Academy of Sciences at St Petersburg, and of the Lincei at Rome, the Cartel communicating with the Academy of Sciences at Berlin.[34] These proposals were approved by the individual members of the Cartel,[35] and informal movement towards expansion of the membership began.

In the light of what was to happen later, we must emphasise again that all the discussion of membership was expressed in terms of Academies as autonomous bodies, not in terms of governments or nations. This distinction is of considerable importance in appreciating the significance of several stages of development of IRC and of ICSU, as national membership and organisational membership developed side by side. If the

Royal Society appears to stand out in this account there is this good reason: the Royal Society was answerable only to itself. It was distinct in this important way. All the other academies had been set by governmental or royal authority. The Royal Society, in spite of its name, began as and remained a private body.

Attitudes towards the character of science change. At this stage the science being considered was, on the whole, purely academic. To illustrate this outlook: the Council of the Royal Society had other business at the same meeting as it received the report of its Secretary, Arthur Schuster, on the Academy proposal. Under one item it was resolved that it did not wish to appoint delegates to an International Congress of Applied Chemistry to be held in Vienna in 1898, because it was of 'too technical a nature'. The scope of international concern for science was to be changed by events in the 20th century. The boundaries between pure and applied became blurred.[36]

Within a short time the Academies of Paris and of St Petersburg and the Accademia dei Lincei had said they were ready to join an international association, subject to their approving the detail of its Regulations.

The International Association of Academies (IAA)

The Conference envisaged in the original invitation (one to establish a formal constitution for a new Association) took place in the agreeable spa surroundings of Wiesbaden on 9 October 1899, and produced a substantial set of Proposed Statutes.[37]

The original members were:

(i) Königlich Preussische Akademie der Wissenschaften (Berlin)
(ii) Königliche Gesellschaft der Wissenschaften (Göttingen)
(iii) Königlich Sächsiche Gesellschaft der Wissenschaften (Leipzig)
(iv) Royal Society (London)
(v) Königlich Bayerische Akademie der Wissenschaften (München)
(vi) Académie des Sciences (Paris)
(vii) Kaiserliche Akademie der Wissenschaften (St. Petersburg)
(viii) Reale Accademia dei Lincei (Rome)
(ix) Kaiserliche Akademie der Wissenschaften (Vienna)
(x) National Academy of the USA (Washington)

Nine other academies were invited to join.[38]

The Statutes proposed were mainly concerned with the administrative structure needed to establish and maintain an organisation of such a nature. For example, IV.10 states:

> IV.10 The business of the Association shall be conducted by means of –
> (a) General Assemblies.
> (b) A Council.

This simple, basic provision, of a gathering of representatives of all members, and a managing body drawn from them, survived all the changes of the next ninety years.

One regulation is more important than any other, identifying the main purpose of the body:

III.8 The object of the Association shall be to initiate and otherwise to promote scientific undertakings of general interest, proposed by one or more of the associated Academies, and to facilitate scientific intercourse between different countries.

'Initiate' is a key word. Up to this time the scientific initiatives of each academy had been its own, except where territorial considerations had to be regarded (as in geodetic surveys). Now it was implied that there might emerge from discussion initiatives for common action between groups of different nations as well as between those of the same nation. What had been specific and limited in the case of the International Polar Year of 1882–3 was now to be treated as a general aim. It was to be a long time before any such actions were important, but the germ was visible.

Only some of these academies were, in fact, purely scientific in their interests. 'Wissenschaft' was not the same as the English 'science'. No two countries dealt in the same way with the need to provide for 'natural sciences' on the one hand and the 'humanities' on the other. Some had academies with divisions, while in some countries there existed separate academies for the two areas.

So the regulations included this provision:

V.12 A General Assembly shall consist of two sections, viz., a section of Natural Science and a Literary and Philosophical Section.

V.13 An Academy may send delegates to one or both sections, according to the nature of its sphere of action.

This led to problems in several countries, for example in Great Britain, where (as we shall see) it became necessary to form a new body, the British Academy, which could represent historical and philological pursuits.[39]

The regulations also provided for an 'Acting Academy', an Academy belonging to the place where the next General Assembly was to be held and charged with its organisation. This seems simple and sensible, giving a chance to several countries in turn to exercise a leading role in the Association's affairs. However, this 'passing of the torch' provision became, in part, an obstacle to progress; it soon became clear that, in spite of its moral value, it imposed a certain impermanence, since it was difficult for any shifting authority to hold funds or provide continuity of management. Adherence to the letter of the provision contributed to the breakdown of the Association during the 1914 war (see below for the events of 1916) and a solution was not found until after 1919.

The results of this first constitutional meeting at Wiesbaden, mainly agreement by academies to join, were communicated to the Berlin Academy which provided the preparatory administration. The first formal meeting of the Council of the new International Association of Academies took place at the Académie des Sciences in Paris on 31 July–1 August 1900.

There were present:

H Diels (Berlin: History and Philosophy), W Waldeyer (Berlin: Sciences), Ch Lagrange

(Brussels: Sciences), A Heller (Budapest), S Laache (Christiania), Ed Riecke (Göttingen: Sciences), W His (Saxony: Sciences), E Windisch (Saxony: Philology and History), A W Rücker (Royal Society), P Fürtwängler (Munich: Sciences), F Lindemann (Munich: Sciences), Famintzin (St Petersburg: Sciences), Salemann (St Petersburg: Sciences), L Bodio (Lincei: Italian Councillor of State), G Retzius (Stockholm: Sciences), J M Crafts (Washington: Sciences), von Lang (Vienna: Sciences), Th Gomperz (Vienna: Sciences).

The French Delegates were:

> G Boissier (Académie des Inscriptions et Belles Lettres)
> O Gérard (Académie des Sciences Morales et Politiques)
> J-G Darboux (Permanent Secretary, Académie des Sciences)

The Chair was taken by Jean-Gaston Darboux.[40] Of apologies for absence one was historically unique: the celebrated Italian chemist Cannizzaro could not be present because of the assassination only a few days previously of Umberto I, King of Italy.

Most of the discussion in the main meeting and in the two committees (one for scientific bodies, one for the humanities) was procedural, but a few scientific questions were aired such as a Royal Society proposal for the measurement of an arc of the meridian in Africa (which would require diplomatic agreement between Britain, France, and Germany).[41] All unresolved questions of Statutes were deferred to a General Assembly, to be held in London in 1902, with the Royal Society as 'Acting Academy', being host.

'Science' and 'Wissenschaft'

Great Britain had a special problem. The Association considered itself to represent the world of learning at large and so was open to delegates representing the humanities as well as the sciences. In Great Britain the Royal Society had enjoyed acknowledged leadership in scientific matters for a very long time and so took on that role in the Association negotiations. However, in its corporate wisdom it realised that there was a lack of conformity and that there should exist a British body playing an independent parallel humanities role. During the period of preparation for the 1902 General Assembly, the Royal Society took part in the preparatory work aimed at the creation of a 'British Academy' which should have independent membership of the Association, representing British interest in the humanities. The Royal Society tactfully drew back when the success of the plan seemed assured. The British Academy came into existence by Royal Charter dated 8 August 1902.

(This problem highlights the wide representation of *Wissenschaft* that characterised the Association and made it different from its successors, which were more sharply focussed on the natural sciences. It is interesting to note that in the 1990s international co-operation between natural and social scientists became an issue that ICSU had to address).

The early history of the IAA points many lessons for the future. Different kinds and levels of Association may co-exist or may conflict. The Association did not replace the German Cartel, which continued to function during the lifetime of the Association. This was to be important in the politics of the war-time period.[42]

London

In early 1904 we see signs of the kind of action which was to be one of the main justifications for the existence of an effective international body: moves were made on behalf of the International Geological Congress for co-operation through the International Association of Academies between interested bodies such as the International Seismological Association and the International Geodetic Association, for the establishment of a world system of seismological and geological observation stations. This was debated at the General Assembly, the first effective meeting of the International Association of Academies (other than for administrative purposes). It took place in London in Whitsun week (beginning 14 May) in 1904. Among the scientific topics discussed were:

> The International Catalogue of Scientific Literature
> Seismology: the proposal for an International Seismological Congress arising out of the discussions referred to above.
> Terrestrial magnetism.
> Geodetics: request to the International Geodetic Association to promote study of mountain levels and gravity measurements.
> Geodetic arc measurements in Africa.
> Atmospheric electricity
> Anatomy of the brain: proposal to create a new organisation.
> International recognition of the Marey Institute.

One of the most weighty of the proposals in the humanities was that of the publication of the works of Leibnitz.

The attention paid to seismology was one of the reasons for the interest shown by Japan which applied (through the Tokyo Academy) to join the Association, in time to take part in the General Assembly of 1907, which was scheduled to take place in Vienna.

Vienna

At the meeting which was held in May 1907, seismology continued to be given a good deal of attention. By now the African survey of the 30th Meridian was approaching the limits of Franco-British territory, and it was confidently hoped that Germany would co-operate in allowing the continuation of the survey to the point where it would be profitable to seek the co-operation of the Egyptian authorities. The International Association of Academies was, of course, not contributing to the scientific content of this exercise: but rather acting as a facilitator and as an agent of more or less diplomatic communication. This could also be said of the efforts to enlist governmental support for the establishment of new meteorological stations in high Polar latitudes, and on islands remote but well placed for observation.

Rome

The General Assembly of 1910 met in Rome, where the question was again raised of a domicile for the Association, a question which was coupled with the desirability of the Association holding funds of its own. It may seem strange at the present time, when money matters and the acquisition and control of funds are a constant preoccupation of all learned bodies, to think of any predecessor considering the holding of funds to be beneath its dignity. The feeling was not unanimous, by any means, some smaller countries appreciating that they could not, for example, host substantial meetings without being able to call on some central fund for organisation and hospitality. Funds could only be held if there were a legal domicile. A legal domicile would only be of benefit to the Association if it existed in a place where the Association could benefit from the local financial laws in the management of funds. During the meetings in Rome, attention was drawn to one fact which has a bearing on the future of the Association and its successor bodies, namely that Belgium had enacted a law providing special facilities for International Associations. However the issue remained unresolved until the IAA had been replaced by the International Research Council after 1919.

St Petersburg

At the General Assembly which met in St Petersburg in the week beginning 12 May 1913 the range of topics discussed was widening, but their range enables one to see both strengths and weaknesses of the Association. Some of the subjects discussed had become perennial (e.g. terrestrial magnetic surveys) and the discussions ineffectual. Some were debated because they had territorial implications which required the Association to play its useful diplomatic role (e.g. investigators of volcanoes had often to seek permission to enter foreign countries).

The discussion which had opened in Rome on the rationalisation of the terminology, identification, and description of colour continued. It was clearly of great importance in the physiology of vision and in many branches of physics, but was never to come to a satisfactory conclusion within the IAA because it was overtaken by the eventual establishment of industrially useful standards.

Another topic considered by the Assembly, the simplification and unification of the calendar, has deeper roots than are obvious in daily life. The calendar itself reflects the condition of the earth's motions so the measurement of terrestrial and civil time is tied to astronomical time. The conduct of economic and social affairs is so tied to the calendar that the discussion of its scientific basis comes very close to the public domain.

Two subjects were, so to speak, internalist. One was the international production of a catalogue of scientific literature. Efforts at establishing a common reference to the literature of science had been made for a generation or more, with some academies like the Royal Society speaking from experience of their own efforts in making useful lists, others like the Berlin Academy urging the adoption of simple reference systems which yet others believed would be inadequate. All the time this was going on the mass of scientific papers was expanding out of the reach of any system so far devised. There were, in any case, internationally, too few bibliographers at work and too little agreement about bibliographical methods. Separately from the International Association of Academies, the librarians were elaborating their own approach to the flexible systems of dealing with an expansion which they were able to foresee and analyse better than the scholars they served.[43]

In one area the IAA foresaw a need and did a good deal towards meeting it. Embedded in the literature was an enormous mass of data on physical and chemical constants, values of measures sufficiently well authenticated by repetition and verification for it to be possible to incorporate them in acceptable tables of universal relevance. The publication of these International Critical Tables was an enterprise that required international support of the kind the IAA was ambitious to give.

It was decided to hold the next meeting of the Assembly in Berlin in 1916, and the Royal Prussian Academy of Sciences began to correspond about arrangements, but no one, so far as one can judge now, realised that the St Petersburg meeting had been the last.

In August 1914 war broke out, and every aspect of international relations changed profoundly. Two opposing camps were set up. On both sides mature scientists put their services at the disposal of governments and the military. On both sides promising young scientists died before their time. On both sides scientists refused merely to wonder what might come of it all, but set their minds and energies to making plans for an uncertain future.

Separate from both of the belligerent groups other nations maintained several kinds of neutrality, sometimes truly detached, sometimes coloured by cautious sympathies. The Berlin Academy still had the status of 'Directing Academy' of IAA even though it could not 'act' in the sense of organising an international meeting. In an effort to keep the Association from being destroyed by the war, the Berlin Academy approached the Dutch Academy to ask it to take over responsibility for keeping the Association going. This was reasonable in principle, but proved to be unacceptable because the German submission reserved the right to regain control of the next effective General Assembly whenever it might be held. The proposal died, mainly through objections from the French. Perhaps there was thus lost an opportunity to provide an example of continuity of aim transcending the disturbances of war, but no doubt such a thought would have been more pious than practical.

It was not only the war that gradually diminished the vitality of the academies. Some of the International Associations lapsed by reaching the end of the lives envisaged in their founding conventions. The German proposal of a temporary Dutch domicile might have provided a useful experiment, but it was not to be, and for a time authority was divided into many parts. The Germans tried to keep up some lines of communication and work, but they lost their wider international character, being limited to keeping up the structure of the Cartel and forming some new bureaux dealing with affairs in the territory of the Central Powers.

In each country the national academies kept up activities in support of their countries' war effort so far as seemed possible. The St Petersburg Academy dropped out of the picture with the Russian Revolution, and responsibility for Russian interests was eventually claimed (rather hopelessly) by a minority group of emigré Russians. It was to be a long time before Russia came back into the picture. The loss was great. Men like the geodesist Prince Galitzin[44] had contributed richly to the common scientific cause.

German work on the International Catalogue of Scientific Literature went on into 1915 but then died out. In Great Britain the work on it slowed down, to come to an eventual end just after the war. There existed many joint activities such as those in Geodesy which were governed by international agreements, the status of which both during and after hostilities were quite uncertain. The work on an arc of the meridian running through a long stretch of Africa divided between Anglo-French and German controlled territories was obviously no longer feasible.

The practical weakness of the Association was made clear by its war-time experiences. It belonged nowhere in particular, and possessed no resources of its own. Its strength lay in the correctness of the initial beliefs of it founders, that international co-operation in science was not only desirable but should have definite, identifiable form. That belief persisted, and efforts were made to restore its fruitfulness.

A turning-point

Persons make policies and persons do deeds, but no one is irreplaceable. In 1917 Jean-Gaston Darboux, Secrétaire Perpetuel de l'Académie des Sciences, died. He was succeeded by Charles Emile Picard, who was slowly to establish a close relation with Arthur Schuster, secretary of the Royal Society. They became, with George Ellery Hale,[45] Foreign Secretary of the National Academy of Science, the architects of the new structure which began to take shape during the war. Hale had taken a vigorous interest in the war, urging United States involvement. In addition to its military significance he foresaw the possible consequences of such involvement for the whole world scientific community, not just for that in the United States itself. He was ready to join in European discussions as soon the United States entered the war.

In 1917, an initiative by the Royal Society marked a turn in affairs. An invitation was sent to the Académie des Sciences to join discussions on the future of international scientific organisations. The Cartel concept, it is plain, had been lost with the breakdown of relations with those countries for which it had originally had so much significance. A fresh start was needed.

Some issues which were discussed early in 1918 included:

The resumption of international meetings: should these include representatives of the
Central Powers (the group of German-speaking powers and their allies)?
Should there be continued recognition of 'Central Bureaux' which had been estab-
lished in Germany for the interchange of scientific information?

And most important:

The future of the international organisation of science after the war, should it be
decided that the IAA could not continue in its present form.

The new beginning: from IAA to IRC

Some of these considerations remained academic during hostilities, but some informal
action began. Political sensitivity was always present. There was some apprehension by
the Italians, represented by the Accademia dei Lincei, that they had been improperly
left out, and there had to be an exchange of carefully worded letters.[46] However, diplomatic calm prevailed.

The suggestion for an initial Anglo-French exchange of views remained effective, but
communication with the United States was soon seen to be essential. The British and
French could communicate with each other fairly easily and wanted to get on and make
constructive plans, but it was not so easy with the Americans. The main USA objection to haste centred on the undesirability of taking action before the terms of a peace
settlement had been arrived at, and, since there was as yet no peace, the Anglo-French
conversations continued on their own.[47] It was common ground that none of the principal parties was yet ready to admit even neutral powers to a new conference, should
one be convened, let alone ex-opponents.

Arthur Schuster had no illusions about the effectiveness of the International
Association of Academies. He urged that any new body should have more authority,
wider vision, and power to do more active positive scientific work.[48] A preliminary
consultative meeting was convened for late 1918, for representatives of the Royal
Society, the Académie des Sciences, the Accademia dei Lincei, the National
Academy of Sciences, invitations also being sent to representatives of the Tokyo
and Brussels Academies. A preliminary report[49] of a committee of the Royal Society
set up to prepare for the conference gave a useful analysis of the factors to be
borne in mind when considering the formation and value of any international
organisation.

The prospect of progress in the resumption of good international relations in science
at the end of the war might seem to have been sound if goodwill had been the only
emotion at work, but however strong might be the feeling for the universality of science,
there was another factor. The war had engendered a bitterness unparalleled in modern
history. The French in particular urged that the Central Powers and their allies should
be excluded from any new organisations, and their fervour carried the allies with them.

A conference met in Paris in December 1918, and it was decided to create a new organisation. For it to come into full effect, a certain number of countries had to signify adherence. The following were the first to do so:

> Académie des Sciences, Paris
> Reale Accademia dei Lincei, Rome
> Royal Society, London
> National Academy of Sciences, Washington
> l'Académie Royale de Belgique

When this had been done the founding convention could come into force, which it did on 1 January 1920. It was agreed that it should remain in force until 1931 when its future would be considered. (This is an example of a 'sunset clause' so frequently included in the terms of reference of scientific projects and programmes. In this case it was to trigger the emergence of ICSU). The new body was to be known as the International Research Council, Conseil International de Recherches.

Chapter 2
International Research Council (1919–1931)

The condition of science in the 1920s

It would be possible to give some account of the International Association of Academies without referring to very much scientific detail because it was, as its name implies, a matter of academies and meetings of officers of academies. The formation of its successor body, the International Research Council, also took place against a historical background of national politics (friendship, enmity, neutrality), of academic foresight and academic rivalry, all of which had affected the International Association of Academies, its formation and decline, so one might expect to see little difference in their styles of development.

However, when we turn to later events, after 1918, we need to take a wider view. To the background of the early years of IAA there was now added economic confusion, coupled with material and technological progress. Some writers have concentrated on describing the development of IAA and IRC without making much reference to the science involved. Certainly the world view of internationalism in many fields changed considerably in the period we are looking at. Equally certainly many of the personal attitudes of men involved in scientific organisation were coloured and motivated by political considerations. The League of Nations lay at the centre of a picture full of detail, but the transformation which took place in the international scientific organisation scene after 1918 cannot really be understood without reference to the activity of the times in science itself. This had been as revolutionary as anything in politics.

At the outbreak of war, both sides had found themselves short of some essential products of science-based industry. When the war ended, there had developed in some quarters a consciousness of the reliance of national economies on applied science which, although it might vary, was never to fade entirely.

Admittedly all scientific change has appeared remarkable to the contemporary observer.[1] Every public event,[2] or notable date generates a literature of surveys of the state of science,[3] from which we generally get the impression that the writer thought he was living in a time of accelerating revolution. This is all relative: Einstein can hardly have shocked the world more than did Galileo, or Frank Whittle more than James Watt. But change does come about, and its scope can be more important than its speed. The creation of the IRC took as long as the creation of the International Association of Academies, but its potential was much greater, because its structure was more clearly related to the conduct of science than that of the IAA. So we need to look at the state of science at large around the end of the 1914–18 war and soon after.

The physical sciences had progressed in the 19th century to a point at which some commentators asserted that all the major problems had been solved and that there was need only for some tidying up. This view was totally confounded by events in the period 1890 to 1914. A small selection is still enough to astonish. There were the discoveries in radioactivity and radiation. The atom, so far from being one of Newton's 'small, massy, hard, impenetrable particles', had been shown to be nothing of the sort but to have a structure. The discovery of the inert gases of the atmosphere had led to much rethinking of the range of chemistry, the basis of which was being transformed by theories of the structure of the atom providing new theories of the structure of molecules, and of the macro-structure of crystals and metals. Chemistry was becoming one of the most powerful allies of medicine, with a few new drugs like aspirin and salvarsan showing what could be done in areas considered intractable.

The experimental demonstration of the physical validity of the mathematical theory of electromagnetic radiation had given rise to radiotelegraphy, a novel means of long-distance communication. In the technological field, the discoveries and inventions were equally remarkable: the turbine had come into widespread use as an alternative to the reciprocating engine. The powered flight of machines heavier than air had been established on a routine basis. Perhaps most influential was the development of electric power: the means by which energy could be utilised at a great distance from the point of its generation.

The astronomers gazed on and photographed a quite different universe from that which had occupied their forebears of only a generation before. Larger and larger telescopes were being built, the great 100 inch telescope at Mount Wilson coming into commission just as the post-war discussions were getting under way. The spectroscope had been recruited into the service of astronomy soon after its development for chemical purposes in the mid-nineteenth century, but now it was being used to measure details of stellar behaviour and structure that some would hitherto have supposed impossible.[4] One striking link between sciences apparently remote from each other was the discovery that the element supposed to exist in the sun, helium, existed on earth and could be isolated.[5] This demonstration that the earth was of the same kind of material as the heavenly bodies was as disturbing in its way as Darwin's placing of man in the zoological world. It was also becoming clear that the universe was not an empty space beyond the Milky Way, but a population of galaxies of like size. It was also clear that the source of the energy of the sun had to be sought in the new physics that was developing so rapidly. Coupled with this were questions of age and duration, of the earth, the sun, and the stars themselves.

Medicine was absorbing many benefits from biological sciences and their techniques, as well as improving its own methods. To give only one example, the techniques of histology were making the microscope an ever more powerful means of detecting and diagnosing disease, as well as of studying the nature and behaviour of tissues and micro-organisms in health and disease. Radiography was giving the cardiac and orthopaedic surgeons a literally new view of the body.

Needless to say, the growth in each of these branches of science–under the umbrella of national academies–promoted specialisation and concomitant emergence of national societies and associations each devoted to one discipline. Thus the professional scientific world was changing fast, and it had to find new ways of organising its inner lines of communication. The International Association of Academies was not adequate to meet the ambitions of those who wanted to make a unified community of science, and some new structure was obviously needed.

It may seem absurd now, when war is seen as a total suspension of normality in many types of relationship, to contemplate the continuation of academic communication between military combatants. But in 1914 there were, as we saw, those who thought that the position of the International Association of Academies could be stabilised, albeit in a form modified to meet new requirements. This fine hope failed, but total war did not lead quite to total disillusionment.

Collaboration: meeting in hope

During the 1914–18 war the nature of international collaboration in science in a hoped-for peace was seriously discussed. What might be the shape of international science when a peace eventually came? The question was asked even though practical scientific work was masked by, or subsumed in, the over-riding needs of the military. Although the condition of political and social thought was to a great extent, inevitably, coloured by ideas of conflict and opposition, there had been running for many decades a deep undercurrent of thought about the nature of international political collaboration, even of the creation of some overriding authority that would keep the peace and promote the general good. The League of Nations did not spring fully armed from the head of President Wilson, although he took a leading part in bringing it into existence. It was born after a long gestation.[6] Those who thought that part of that general good was the promotion of science had no difficulty in discussing plans for a new scientific organisation.

Belief in a possible unification of effort was encouraged by the fact that in most investigations a chemist, physicist, biologist, or medical man might feel at home with like-minded colleagues of any university or any country. One effect of this, which had an important influence in the matters with which we are concerned here, was that scientific men on all sides had long felt that there was an independent unity of discourse about the nature of the external world to which they could all eventually resort at the end of hostilities. The administrative issues were subordinate to the intellectual issues. However, men act by emotion as well as by logic, and war is an emotional business for everyone.[7] We must not be surprised, therefore, to find the plans made and the actions taken falling somewhat short of a moral ideal and the medieval dream.

Be that as it may, an effective initiative grew out of the communication between a few individuals. The 1917 correspondence which, as we have noted, had been set up between several men in Great Britain, France, and the United States was not intergovernmental, but deeply personal. The leading figures in this action (the American astronomer, George Ellery Hale, Secretary of the United States National Academy of Sciences,[8] and the British physicist Arthur Schuster[9]) shared an enthusiasm, and their opinions carried weight. Action was taken by the principal national academies of the United Kingdom, France, Italy, Belgium and the United States, which, encouraged by the consultative meeting on 8 May referred to in the previous chapter, held a conference in London in October 1918. It adopted the following resolutions:

> 1 That it is desirable that the nations at war with the Central Powers withdraw from the existing Conventions relating to International Scientific Associations in accordance with the Statutes or Regulations of such conventions respectively, as soon as circumstances permit.
> 2 That new associations, deemed to be useful to the progress of science and its applications, be established without delay by the nations at war with the Central Powers, with the eventual co-operation of neutral nations.

At that time, any discussion was bound to be overshadowed by the condition of war, some parts necessarily being expressed in terms of the opposing sides 'Allies' and 'Central Powers'. This opposition was to persist for some years (and to find a reflection in the events of the period after the Second World War). A decision was made to form a Council which would co-ordinate international activity in the various branches of science, and would do it in such a way as to promote the formation of international bodies, associations, or unions, devoted to distinct branches of science or collaborative projects. This concept was to provide a firm outline to the pattern of international activity which has persisted to the present time.

In November 1918, a further conference (a continuation of the London conference) in Paris decided to extend the membership and to invite several neutral nations to be represented on the new body. They were Denmark, Spain, Monaco, Norway, the Netherlands, Sweden, Switzerland, Czechoslovakia, Finland. This list reflects at once the political changes which followed the war and which were to be a feature of international politics down to our present day: the concept of the nation-state being a constant element in negotiation and co-operation and, unhappily, in antagonism.

The conference had to resolve several problems, some already suggested in the 1918 London resolution. One was how to disentangle the relations already existing between old associations, so as to leave a clear field for the creation of a new body. Another was the perennial problem of membership. The old nations were changing. The British Empire had developed the concept of Dominion status, with what had been colonies now becoming independent nations with distinctive cultures, ambitions, and, most important in this context, separate scientific establishments with some degree of independence.[10]

An element of a new character came on the international scene: the League of Nations. Early in the war discussions had begun about the hope of forming some association of nations which would diminish the prospect of further war. There was strong United States influence in this movement, but eventually the USA did not become a member of the body which was set up. The League of Nations was constituted officially, after much preliminary negotiation, on 10 January 1920. (Its eventual formal dissolution was to be voted on 18 April 1946).[11]

There were new alignments in Europe, with some new nations being formed out of surviving fragments of old cultural elements, Czechoslovakia being the most prominent. Finland had been part of a Russia, the future of which was still uncertain as one form of governing party succeeded another. The Balkans, where the war had been triggered off, had not yet settled down. Serbia seemed still to survive as an independent unit, but it was not to survive in the same form for long.[12]

The new scientific body needed a strong name; it had been decided at the November 1918 meeting that it should be called the International Research Council, Conseil International de Recherches, a name it retained until changed circumstances dictated a further change of name in 1931. The International Research Council came into formal existence with the approval of a constitution based on the draft Statutes proposed by the provisional Executive Committee. This was done at a Constitutive Assembly held in the Palais des Académies, Brussels 18–28 July 1919. The opening session was honoured by the presence of H M The King of the Belgians. King Albert had only lately been restored to his position in his liberated country. The choice of Brussels for an inaugural meeting and this royal presence underlined the deep nationalistic feeling that was to be a feature of painful and protracted negotiations over the comprehensiveness of IRC membership for a long time ahead.

The effective business of the Assembly was under the Presidency of M Gravis, Vice-Directeur de la Classe des Sciences de l'Académie royale de Belgique.[13] The objects of the IRC International Research Council were simple and realistic:

1 To co-ordinate international efforts in the different branches of science and its applications.
2 To initiate the formation of international Associations or Unions deemed to be useful to the progress of science, in accordance with Article 2 of the resolution adopted at the Conference of London, October 1918.
3 To direct international scientific activity in subjects which do not fall within the purview of any existing international associations.
4 To enter through the proper channels into relations with the governments of the countries adhering to the IRC in order to promote investigations falling within the competence of the Council.

As shown at the beginning of this chapter, the rapid development of science was leading to increasing specialisation and formation of national professional organisations. It is quite striking that the objectives of the IRC reflect on an international scale what was taking place nationally: promotion of professional interaction within one discipline through the formation of Unions, and at the same time provision of a framework for Union interaction under the umbrella of the collective national academies.

Since the objectives speak of 'countries adhering to the IRC', it must be said that the definition of 'national membership' has evolved over the years under the influence of two considerations: preserving non-governmental status and avoiding political issues resulting from territorial disputes in such a way that, as much as possible, academies from the whole world would be included. In 1993 the ICSU Statutes defined a 'National Scientific Member' as 'a scientific academy, research council, scientific institution or association of such institutions' while 'institutions effectively representing the range of scientific activities in a definite territory may be accepted as National Scientific members provided they can be listed under a name that will avoid any misunderstanding about the territory represented'.

We shall have occasion to return to this later. Meanwhile the reader should realise that for indicating 'National Scientific Membership' in this book the old terms like countries, nations and national members are being used according to the usage of the time period under discussion.

A problem which was solved by not being too rigid was that of the nature of an adhering national body. The United States was represented by its National Academy of Sciences, which proposed that each country should set up a National Research Council. In the United States a National Research Council had been created to examine war-time scientific needs and problems, but its impressive achievements led to its being given a permanent position in United States official scientific life.[14] The suggestion was intended to ease the way to membership for those countries which had an academic structure different from those in countries like the United Kingdom or France which had a long history of scientific academies.

Thus it was possible to declare that the representative body for a nation could be its principal scientific Academy, a National Research Council or equivalent, or some national institution which on its own or by association within the nation fulfilled analogous functions. The adherent might even be some office of the government of the country, so long as that office existed for a scientific purpose. (This provision was to be eliminated after the Second World War, so that ICSU could maintain a strictly non-governmental status).

Opposing powers

All this was, formally, neat and tidy, but the discussions must be viewed against their background, which was that of the physical horrors of war and the political, economic, and emotional aftermath of war.

Membership of the Council and of the Unions which the Council aimed to co-ordinate or to bring newly into being was restricted in the first instance to the 'Allied Powers'. The 'Central Powers' were excluded. The London conference of 1918 had envisaged the membership of neutral countries, but in the first instance the countries represented in the IRC Constitutive Assembly were: Belgium, Brazil, Australia, Canada, the United States, the United Kingdom,[15] France, Greece, Italy, Japan, New Zealand, Poland, Portugal, Rumania, South Africa, Serbia.[16] These countries had all been belligerents on the side of the Allied Powers or closely associated with them.

There was a further provision:

> After a Union has been formed, nations not included in the above list, but fulfilling the conditions of Article 1 of the resolutions of the Conference of London, may be admitted, either at their own request, or on the proposal of one of the countries already belonging to the Union. [This Article 1 was that requiring withdrawal from existing conventions.]

The initial membership conditions of both Council and Unions thus excluded neutrals, at least for the time being.

In most organisations, the first meeting of interested parties is quite different in character from general meetings convened later on when a workable constitution has been worked out. This first constitutive assembly of 1919 had very little to do with science and much to do with pride and *amour propre*.[17]

The scientific disciplines were to reside in the Unions (potential as yet but ready to be formed), and, although relations between Unions and IRC would naturally be an important matter, they were never sufficiently characterised for IRC itself to see at all clearly its own role. It was agreed that the Statutes of each member Union should contain certain provisions which reflected the character of IRC, but there was an oversight: there was yet no condition that a country should be a member of IRC before being eligible for membership of a Union. In other respects the Statutes were sometimes difficult to interpret, as they would affect both IRC members and Unions. This seems to be one reason why a spirit of uncertainty at the formative stage prompted the members to agree to the first phase of the life of IRC being limited to ten years, after which a new constitutive assembly would be held to discuss the future. From then on the date 1931 hung over every discussion of forward planning.

IRC: First General Assembly: immediate problems

The first General Assembly (Brussels, 18–28 July 1919) consisted only of national delegates, no Union representation having yet come into effect. Voting was on a scale of voices, (1 to 5) depending on population. Contributions due were scheduled at a range of 1 to 8, also on a population basis. (Populations of dependent territories could be counted.)

The administrative structure agreed was fairly simple. An Executive Committee would consist of five members elected by the Assembly together with a Bureau (President, two Vice-Presidents and a General Secretary, all elected by the General Assembly) for the management of business.

Important and wise appointments were made to this Executive Committee: as President, Charles Emile Picard, the distinguished French mathematician, Secrétaire Perpétuel de l'Académie des Sciences, Paris, and, as Secretary-General, Arthur Schuster, Secretary of the Royal Society, London. The sum of experience of these two men was formidable, and the personal relation between them became cordial.[18] With them, as Vice-Presidents were V Volterra of the Lincei and G Lecointe of the Académie royale de Belgique. George Ellery Hale, Foreign Secretary of the National Academy of Sciences of the USA, was an additional 'ordinary member' although a less 'ordinary' member than this influential astronomer would be hard to find.

While the preparatory work of this Committee was going on, there was a good deal of supportive activity in related fields, with co-ordinating effort in such fields as Astronomy, Geodesy and Geophysics, and Chemistry. These efforts were to result in the consolidation or foundation of the kind of body which was to be a characteristic member-body of the new International Research Council, each one having the word *Union* in its title. Scientific Unions had been in existence for a long time, the word *union* being a useful alternative to *association* or *society*. But now it began to take on a distinct significance, as a form of scientific association which existed to initiate and develop useful agreed action. It may seem odd that IRC and ICSU managed to develop without a restrictive definition of one of the most important elements in their constitutions, but develop they did.

This word *union* is important, even though it is not easy to offer a definition which will survive, unaltered, translation into the many languages of ICSU members. Later on the existence of the specific term in the title of a certain set of bodies was to be helpful in making clear the significance of other kinds of term (even the familiar 'committee') for identifying bodies created for other kinds of purpose and for showing some degree of hierarchical subordination.

Draft Statutes for the Council itself and for International Unions of Astronomy, and of Geodesy and Geophysics were adopted at a meeting of the provisional Executive Committee in Paris 20–24 May 1919.

For a simple reason, the first Unions to be created were for Geodesy and Geophysics (IUGG) and for Astronomy: their basic studies had always required international action so international 'union' came naturally.

The form of association in the physical sciences (chemistry and physics) still presented some problems, but eventually satisfactory systems of organisation (IUPAC and IUPAP) were devised.[19] One other Union (IUBS) was constituted at the same meeting, but did not become a full member of IRC until 1925.

Chemistry is a complex subject for which many bodies already existed, some dating back to quite early in the 19th century. An interallied conference of pure and applied chemistry met in Paris in April 1919 and agreed on preliminary proposals for a confederation. The title of the new chemical organisation, International Union of Chemistry, sometimes appears as International Union of Pure and Applied Chemistry, revealing a new attitude among the academics: the word *Applied* marks the recognition of the need to study and influence the practical side of scientific work in the community at large.

In one respect the founders of IRC decided not to be caught out like the International Association of Academies in which there had been some discussion of an administrative base, a legal domicile. It was all very well for periodic meetings to be

held at different centres, and even for some administration to be in the hands of officers of different nationalities from time to time, but there ought to be some element of
'a local habitation and a name', a legal existence subject to some national law. IAA saw
this need but did not meet it. From its beginning, IRC did better. It was agreed that
there should be a legal domicile, and that it should be in Brussels. The Convention
establishing the Council should come into force on 1 January 1920 and should so
remain until 31 December 1931, when, subject to the assent of adhering countries, it
should continue for a further 12 years. The style and manner of these provisions were
sufficiently flexible for this to be a sound basis for action for the first 12 years, but not
only did the revision envisaged for 1931 turn out to be radical and fundamental, the
second period (1931 onwards) was to be much more than a steady continuation: it was
to run into world events of a kind that some might have begun to fear, but the form of
which that could not have been foreseen.

At the beginning there was no fixed scheme for the overall scope of Unions. Strongest
were the astronomers, understandably so since they had the longest history of international activity. A form of the International Astronomical Union already existed, with
statutes which, when settled in 1919, offered a model for other Unions. However, what
suits the logic of astronomical activity would not necessarily suit that of the chemists
or the biologists, so no rules were laid down about internal structure of Unions,
although the General Assembly did foresee the need for fostering inter-Union
collaboration. They adopted this resolution:

> The Executive Committee shall have power to organize researches requiring the co-
> operation of several branches of science, if submitted by one of the National Research
> Councils. It may for this purpose nominate Special Committees or refer the subject of
> the research to the appropriate Unions or proceed in any other convenient manner.

Each Union had its own history and character. The astronomers could look back on
a long, long past. The radio scientists could only look back less than one generation.
The Union Radio Scientifique Internationale, URSI, (which was formed out of an
existing body) was in the hands of men some of whom had been born before the date
of the first experiments exhibiting a radio phenomenon had been performed. For a
decade or so there were no additions to the original founder Unions but after 1945
there were many examples of this kind of 'new science' development. The International
Union of Geodesy and Geophysics already had that long history which we have
referred to, looking back as it did to the International Geodesy Association.[20] The
International Union of Pure and Applied Chemistry also had its antecedents in many
international meetings on chemical topics and a successful effort to form an
International Association of Chemical Societies. A new movement leading to IUPAC
began in 1919. An International Union of Mathematics was of more recent development, but it too seemed to be ready for membership of IRC. (It was, in the event, short-
lived. It was dissolved in 1932, but was resurrected in 1952.[21]) IRC, that is to say, was
based on the 'old' sciences. The 'new' sciences, like X-ray crystallography, hardly
existed in 1919, but were developing as IRC settled down.

The long titles of IUPAC and IUPAP (International Union of Pure and Applied Physics) seem to have envisaged separate areas of interest within the Union, of a kind not so obvious in other Unions, but most of them quickly set up quite elaborate internal arrangements which would serve sectional activities. Thus IUGG set up six sections, on geodesy, seismology, meteorology, terrestrial magnetism and electricity, physical oceanography, vulcanology. All had existed in some form or another for a long time, but now they were given recognised and effective links with each other.

The International Union of Biological Sciences, IUBS, began with six sections, each of broad scope: general biology, physiology, zoology, botany, medical science, applied biology. IAU saw things differently and proposed 30 different commissions charged with the study of particular problems rather than covering interest areas.

With the formation of Unions for the physical sciences, for the biological sciences, and for geography, the coverage of science at large was pretty well representative although not entirely comprehensive (geology was not represented as a Union until 1961, perhaps because many of its needs had been well-served since 1878 by the succession of International Congresses).

In the first three years of its existence, IRC had thus succeeded in doing at least one thing of fundamental importance: it had created a forum in which scientists from different unions could meet to identify common problems and aspirations. (This is a real – and continuing – problem: scientists do not necessarily understand each other's technical language, let alone see what is driving the front of research in a remote field). This system promised to complement that dissemination of scientific information by publication which was (and still remains) an essential means of scientific communication. (What difference electronic communication will make to scientific practice as distinct from the recording of its agreed results remains to be seen.) In particular it promoted that testing of tentative theories against the critique of peers which is so essential to scientific progress

Second General Assembly: politics

At the Second General Assembly, held at the Palais des Académies, Brussels 25–29 July 1922, Picard was able to report the consolidation of the position of the Council. The scope of membership was now extended to admit Unions as members. The Executive Committee was enlarged to admit one representative of each member Union.

It is a reflection of a pattern of world politics of the time that provision was also made to admit as members diplomatic protectorates of member countries. The concept has passed away, but this is a reminder that we must take care not to think of the political pressures on IRC as if they were just like those we see operating today.

There remained two overriding, substantial problems. What was to be done about the Central Powers? Had the IRC yet arrived at a satisfactory relationship between the Council and the Unions, and a mutually agreeable basis for conditions of membership of each? It is interesting to compare the different bases for these two problems. The

Central Powers problem was a hangover from the past, not just the recent past of conflict, but a long past of nations warring against nations, the typical and accepted components of the world structure. The Union membership problem, on the other hand, was an indicator of a change in world structure, not so much in political power as in world influence, an influence which had no boundaries because, at its best, science has no boundaries. (The cynic will want to point to examples of greed, ambition, dishonesty, corruption in the scientific world, and all we can do is regret them, but we are also entitled to assert that they are irrelevant to the main stream.)

The passions of the war were still felt, and the Central Powers were not yet, in most minds, admissible. The neutrals were, however, ready to be admitted and several were. The countries represented at this IRC Second Assembly were: Belgium, Canada, Denmark, Spain, the USA, France, the United Kingdom, Greece, Italy, Japan, Monaco, Norway, The Netherlands, Poland, Portugal, Sweden, Switzerland, Czechoslovakia. It was decided to invite three others to join: Egypt, Peru, and Morocco.

Picard confronted the scepticism about the value of IRC felt in some quarters, by putting this to the Assembly:

> Comme toute entreprise d'un type nouveau, elle [IRC] a rencontré ça et là quelques
> scepticismes. Réunir, fut-ce par un lien assez mince, autour d'un Conseil central, tant
> d'Associations, a paru chimérique à certains, qui préfèrent une indépendance absolue
> de divers organismes scientifiques internationaux. Si courte que soit notre histoire, elle
> paraît montrer que ces craintes ne sont pas fondées. La tutelle du Conseil International
> de Recherches sur les diverses est bien légère. Il y a plus: on peut penser que certains
> Unions ont été heureuses de laisser au Conseil international le soin de prendre des décisions qu'elles ne se souciaient pas de prendre elles-même.[22]

An awkward constitutional question was whether there should be a mandatory Statute requiring that membership of a Union should be allowed to a country only after it had adhered to IRC. Eventually this was formulated as follows:

> That only countries which have adhered to the IRC are entitled to be members of the
> Unions connected with it.

Picard's confidence in the international bonding which united the members of IRC still had its critics. It was not only in the scientific field but throughout the politically conscious world community that in many organisations (IRC was only one among many) internationalism had only been partly achieved, if it still excluded the Central Powers. It was beginning to be thought that there had been no winners in a war that need not have happened, and that discussions of guilt were an impediment to a return to a better world.

Third General Assembly: uncertain prospects

So the Third General Assembly of IRC, in Brussels 7–9 July 1925, brought the issue of the Central Powers into formal debate. Picard used words which must have been deeply moving to anyone who was aware of how the war had destroyed Picard's own family. He might have been forgiven for being irreconcilable, but he said:

> Les événements, dont le monde a été le théâtre il y a quelques années, nous ont rappelé durement des vérités trop souvent oubliées. Vous aurez à voir un jour à quelles conditions et dans quelle mesure il conviendra de jeter un voile sur le passé.[23]

The lead in pleading for relaxation of the restraints was taken by 'neutrals' The Netherlands, Denmark, and Sweden. The Dutch submission was put in a forceful letter by H A Lorentz, on behalf of himself and three colleagues of great distinction, J V Enthoven, H Kamerlingh Onnes, and P Zeeman. Here are a few words from it:

> Il y a des domaines dans lesquels la collaboration avec les pays centraux s'impose; aussi a-t-elle déjà commencé, bien que ce soit souvent par des voies détournées et indirectes. Il existe, en dehors du Conseil des Recherches, des organisations dans lesquelles cette collaboration se pratique régulièrement, et plusieurs commissions émanant de la Société des Nations en ont donné l'exemple.

> Nous estimons que le moment est venu de rendre aux efforts scientifiques le caractère d'universalité qu'ils doivent avoir autant que possible, en vertu de la nature de la science même, et nous pensions qu'on peut faire ce pas sans hésiter.[24]

Australia supported a motion for the striking out of the restrictive clauses from the constitution; so did IUPAC. Switzerland echoed the new note struck by Lorentz: they added a proposal that any country that adhered to the League of Nations might be admitted to the Council. These attitudes, although having a common aim, are interesting in three different ways. The Australian support indicates the growing independence of a British Dominion, expressing a mind of its own. The chemists of IUPAC were more aware, perhaps, than others of the great German contribution to science, not least in their field. And the Swiss proposal reminded IRC (as it should remind us in following through this period) that all these IRC events were taking place against a backdrop of the League of Nations.

The moves to admit those still outside failed, for two reasons. One reason was the continued inability of Belgium yet to draw Picard's veil over the memories of a country which had been the first to suffer the destructive effects of 20th-century warfare. The other was constitutional: a favourable vote failed because it was not achieved with a sufficient majority.

A further proposal by The Netherlands and Danish delegates, that a Union should be allowed to accept into membership a country which had not yet adhered to IRC, was lost (on majority vote) on the grounds that its acceptance would diminish the authority of IRC itself. If we may be allowed the wisdom of hindsight, we may see in this fear of loss of authority a sign of a rigidity in IRC as originally conceived that was eventually to demand a total remodelling of its constitutional philosophy, indeed of its replacement. But before this happened the easement of membership conditions prepared the way.

No other Unions were being formed; one, IMU (International Mathematical Union), even dropped out of IRC, but an increasing number of countries were adhering to IRC. Several of them had no strong record of scientific work, but their adherence can be taken as signifying the realisation among them that science was to be an increasingly important factor in their national and international standing. Moreover, we must constantly remind ourselves that science was not isolated in world culture; there was a great deal of activity, self-generated but also stemming from the corporate ambitions of the League of Nations, in forming other world organisations for cultural interchange. When it came to counting pennies, each nation had to see its science as part of a wider culture, and consider the cost of participation. This widespread consideration of the wider culture was one of the roots of UNESCO, but such a future event was as unpredictable as any other peering into the future. In the 1920s, science, like the humanities, was just beginning to create a new international consciousness, and trying to do so by looking to links between the sciences. The IRC/Unions relationship held out hope.

The passage of time in the life of any organisation is nowhere more strongly marked than in the passing, by retirement or death, of those who have chosen to treat it as important in their lives. We see some organisation seemingly identified with some strong person, and wonder how it can survive when he or she chooses, or is forced, to leave office. So with IRC. At the end of the 1925 Brussels meeting, Hale and Lecointe retired from the Executive Committee for reasons of health (although Hale was to return). Both had been important to IRC. Pelseneer of Belgium and Volterra of Italy took their places, but for a while Schuster was able to continue as Secretary-General.

One cannot ignore the effect of political events when considering the atmosphere in which decisions are taken. The publicity surrounding the signing of the Locarno Treaty in 1926[25] can hardly have failed to create an atmosphere favourable to a change in attitude to former enemies. Be that as it may, on 29 June 1926 at last came the turning-point for which so many had hoped. Picard called an extraordinary meeting of the General Assembly at which the principal business was the deletion from the Statutes of the resolution adopted at the London meeting of October 1919 excluding the Central Powers from membership. This was accepted, and invitations were sent to Germany, Austria, Hungary, and Bulgaria to adhere to IRC.

No executive decision of this sort is ever free from administrative complications, and this was so in the case of Germany. The old history of the German academic community had to be taken into account: there was a representative national academy in the other three countries to which invitations could be addressed, but Germany, in spite of its apparently dominant position, had none. So invitations were sent through the Foreign Office in London to the Academies of Berlin, Göttingen, Leipzig, and Munich. All replied in the same way: they all had to refer the matter to the Verband der deutschen Akademien, which did not have the status of the single national academies found in other countries. Austria said it also relied on its affiliation to the Verband. So no dramatic consequence followed this striking change of policy. Hungary said it would join IRC but could not afford to join Unions. Bulgaria said it was too poor to join anything.

As President of IAU, but speaking for other Unions as well, W de Sitter expressed an opinion which illuminated one of the central problems of the philosophy of IRC, namely that, while Unions might be subordinate to IRC during their formative period, they had to be considered as independent of, although respectful to, IRC in their years of maturity. He used the image of the family and the way the early dependence of children turns into co-operative independence within a family.[26] The 'family' metaphor turned out to be persistent and valuable, and is still used about ICSU today.

Fourth General Assembly: preparation for change

Families change character as the children grow up, and the Fourth General Assembly of IRC, meeting at Brussels on 13 July 1928, began the business of preparing for the ending, due in 1931, of the convention which had launched IRC in the first place.

This General Assembly faced the fact that time had flown and that by the date of the Fifth General Assembly IRC would have to be ready to implement the decision of the First, taken in 1919, to review the whole nature and constitution of IRC, and consider its future, whether it would lapse or be continued in some form.

The Royal Society proposed that a Committee be formed to consider changes (in the Statutes, to the methods of voting, and to the management of financial affairs) and, to help its deliberations, this Committee should consult with the Unions. The Italian delegate (Giannini) thought the Unions should be entirely autonomous and that the IRC should have the function only of co-ordination, but this was not supported. There was support, however, for the idea that the Council should be independent of influence by governments and should have no diplomatic ties.

The freedom of each Union to determine its own Statutes (subject to approval by the IRC Executive Committee) was agreed. A Commission on Statutes was appointed with the formidable Henry Lyons (Secretary of IUGG since 1919) as its Secretary. He was to become a leading figure in IRC/ICSU affairs, as had Schuster who withdrew on health grounds in late 1928.[27]

On 26–27 September 1929 the Commission on Statutes of IRC met in Paris to consider the suggestions sent in by 16 of the adhering organisations, to further consider a draft set of revised Statutes and the observations on them of the Executive Committee and the Executive Committees of each of the Unions.

The Executive Committee of IRC made further comments on 7 July 1930 and the final version of the Draft Statutes was ready for submission to the General Assembly for approval.

IRC/ICSU: transformation scene

The Statutes were approved unanimously and had the effect of dissolving the International Research Council and creating the International Council of Scientific Unions. When IRC closed its Fifth General Assembly, on 11 July 1931, there began on that same day the First General Assembly of ICSU.

Chapter 3
ICSU: beginning and establishment

Formation of ICSU

G M Trevelyan said the historian must never forget his duty to narrate.[1] True enough, but sometimes it is useful to make certain of a starting-point. To understand the long life of ICSU one must spend a little time concentrating on the events of 1931. The most important difference between the beginning of IRC and that of ICSU was this: IRC began as a body of national members. Some Unions became be associated with it one by one but with limited powers. ICSU began as a body with both national adherents and Unions as full members. The IRC membership began with 16 national members with full voting powers and 7 associated Unions. In 1922 Unions were given representation on an Executive Committee. Following a growth in national membership IRC provided ICSU with an initial founding membership of 40 national adherents and 8 Unions, all with voting powers in the General Assembly. Unions had a majority of the votes in the Executive Committee. (The International Mathematical Union withdrew after a short time leaving only 7 Unions until after 1945.)

The decision to reconsider the form of an international science organisation by 1931 had taken beforehand in 1919, by people experienced in the way all organisations change, whether they be families, football teams, or federal governments. Such changes can be trivial adjustments: they can derive from profound changes of principle. In the case of IRC and ICSU the change was profound, an almost total reversal of the relation between ICSU and the Unions. Union subservience changed to equality with the national adherents.

In 1928, the Fourth General Assembly of IRC had, as its main preoccupation, the consideration of whether it should continue, with a few modifications, or whether it should change radically, in which case it would have to determine the nature of its own successor body. That a radical change was needed had, however, already been recognised. On 11 July 1931, the Fifth General Assembly met and converted itself into the First General Assembly of a new body, the International Council of Scientific Unions.[2] The new name recognised the reality of the situation that had slowly developed: in the early days of IRC the parent body had patronised the Unions it had fostered. Bit by bit the Unions had come to help justify the IRC, because of the independent scientific strength they acquired as they matured. Since, at all stages of the IAA and IRC (and ICSU), the national adherents outnumbered the Unions by far (see Appendix 2) and the national members provided the bulk of the annual income, it was logical to compensate for this in the name of the new body as well as, as we shall see, to some extent in the voting rights of the Unions. This emphasised their vital role in an organisation that essentially remained an enterprise of the collective academies of the world.

Each Union covered its possible scope in patches, not comprehensively like a textbook, but more often like the papers in a journal, highlighting projects and particular interests. There was already a long history, often going back to the 19th century, of international congresses devoted to special subjects: chemistry, geology, mathematics, several medical subjects, and others. Each Union had as a central responsibility the maintenance of these successive congresses, but the Union concept offered, in addition, continuity in between. To use a fanciful image, each congress might be a jewel, but a Union provided the string that made them into a necklace. And, of course, the existence of a permanent organisation made it possible to do other important things than hold congresses. A 'Council of Scientific Unions' suggested yet another factor of cohesion.

Picard's introduction, as President of IRC, to the proceedings of the General Assembly of 11 July in Brussels[3] said what was necessary. To translate and summarise:

> In 1919 at a time when a League of Nations was being envisaged, it had not been unexpected that an international scientific body of a new character should be formed, and this had happened rapidly, with an uncomplicated acceptance of IRC and its Statutes in their first form. Many international scientific bodies had previously existed but there were now new opportunities for their co-ordination by IRC. But during the short lifetime of IRC there had been at least one great change in scientific activity: an increasing amount of penetration of one kind of science into another, a fertile source of new discoveries. The character of the unifying body therefore derived more and more from that of the member Unions. The time had come not only to recognise this but to change the Statutes of the unifying body accordingly, so that the driving force was the scientific spirit of the Unions, rather than the national interests of the national members. Nevertheless the constitution and energy of IRC (and its successor ICSU) depended equally on the gradual increase in national concerns for science as factor in cultural and economic life.

(In 1925 when de Sitter had first used the image of a family he also referred to the structure of IRC being like a pyramid with the Unions at the base, which might one day be found to be the wrong way up, as the work of the Unions became more and more important and would deserve to be thought of as the apex, rather than the base of the IRC pyramid.[4])

As examples of the interpenetration of one science into another Picard naturally chose from topics close to his own interests: one was the common ground now being found between astronomy and geodesy. Another (and more significant in its link of pure with applied science) was the way the physics of Hertzian waves had become the foundation stone of radio-communication.[5]

One can multiply examples. Biochemistry was expanding, with new discoveries making themselves felt in many directions. Medicine benefited from new knowledge in the age-old fields of study, but was developing new approaches to health and disease. Agriculture was expanding with the understanding of the behaviour of other influences on plant growth than the familiar fertilizers. The growing science of nutrition was coming to terms with the now well-recognised vitamins.

The boundaries between pure and applied science were becoming more and more indistinct, as, for example, the astronomer, the geographer, and the practical navigator came even closer together than they had been in the past.

This covering of the sciences was by no means uniform, however. The geologists were not yet part of ICSU, even though theirs was a science of universal interest and application. They had been holding geological congresses (and continue to do so) usually every four years, since 1878, but they had not yet sought to establish a Union. (This was not done until 1961.[6])

All the same, the structure seemed firm. Picard's address was expressing the feelings of the whole Assembly, and the Assembly was ready to vote. Before this proceeded, however, one proposal was debated which was typical of the kind of accommodation for which ICSU was to be so useful in years to come. It was asked that a commission be formed to carry out research on problems of population (a subject that was going to preoccupy ICSU time and time again, up till the end of the century). This was considered meritorious, but it was agreed that it was best taken over by the geographers, who would be able to relate it to other relevant matters. (This was reported at Paris in 1934).[7] Note a difference here: it was proposed in 1946 to form an X-ray crystallography section of IUPAC but a separate Union was already being planned on the initiative of W L Bragg. It came into existence soon after.[8]

The matter of population studies was a diversion from the business in hand, but it was given a careful hearing: the Assembly returned to its central preoccupation and proceeded to vote unanimously to start a new life.

Initial composition

The national composition of ICSU reported by the General Secretary was now very different from that of the early IRC. There were 40 national members. This must have looked like steady progress, but nations do not stand still: all governments undergo changes of composition and of policy, of background and of financial resources. Italy, of which the representative academy had been the Lincei, did not appear in the 1931 roster of 40 members. It reappeared in 1932. This appears to have reflected the refusal of V Volterra, who had been the representative of the Lincei, to swear allegiance to Mussolini's Fascist state. Volterra and his Lincei ceased to represent Italy. A new Italian National Research Council was formed, which then took up membership of ICSU.[9]

The first officers were:

President: George Ellery Hale (USA)
First Vice-President: G A Ferrié (France) (who died in February 1932)[10]
Second Vice-President: N E Nörlund (Denmark)[11]
Members: P Pelseneer (Belgium),[12] F A Went (Netherlands)[13]
Secretary General: Sir Henry Lyons[14]

The Unions were as before, eight in number: Astronomy, Biology, Chemistry, Geodesy and Geophysics, Geography, Mathematics, Physics, and URSI (the Radio-Science Union always, by custom, referred to by the initials of its French title). (The Mathematical Union became inactive soon afterwards and was re-created in 1952.)

Notwithstanding the constitutional changes, the scientific work went on as before. There were important continuities: an example is the support for the *International Tables of Constants*, dull work, dating back to 1922, but of endless value, and, like everything else, needing more money as it got bigger and more expensive to produce. The change from IRC to ICSU made no difference to its value or its needs.

Another continuity was that of sections formed by the Unions in the past, which carried on their work under constitutional arrangements satisfactory to each respective Union but not in conflict with the requirements of ICSU. Take two examples of the way in which reports of the Unions under the revised ICSU régime were presented, i.e at the Second General Assembly of ICSU in Brussels (9–13 July 1934).

The report of the International Astronomical Union (IAU) referred to mature work on Lunar Nomenclature which had been initiated early in the century; it was now nearing completion and was expected to be wound up. The International Solar Union (founded in 1904 on the initiative of G E Hale) was still active and extending work into further reaches of the solar spectrum. A new committee had been set up on Spectrophotometry, for which new techniques and new uses were developing. An interesting remark in the IAU report was that some of these committees kept in touch with wider sources of observational results by co-opting, as members without voting rights, astronomers from countries which did not adhere to IAU. In other words, a Union could explore ways of extending its own influence, unhampered by the letter of the law of its own or ICSU Statutes.

The IUGG had a history of the existence of several Associations (Geodesy, Seismology, Meteorology, Hydrology), which continued separate lives but clearly converged, in subject interest, on a common ground. It helped them, in budget and in communication, to act together as one Union, but their autonomy was seen in the sections of the IUGG reports to the General Assembly.

ICSU initial constitution

The constitutional changes from IRC to ICSU looked modest but were radical in effect. At the very beginning, in 1919, representation on the Executive Committee had been only national, but this had been quickly changed to allow the Unions a part in the Executive. This early step reflected the change from a purely national membership of the IAA pattern to a dual membership. The actual personnel of the Executive might not look much different from what it would have been under the IRC procedure (capable people are few). However, the Unions could not only feel that they had participated in choosing them, but that their choice was free of diplomatic imperatives which was important at that time. The idea of the Non-governmental Organisation, which was to be so important in the years after 1945, was now implicit in the ICSU constitution. National membership was by way of a scientific organisation (Academy, Research Council etc.). Only where no academy or similar body existed would a government be represented directly.

When the ICSU constitution was worked out it was agreed that the Unions should have a much more prominent place than they had in the IRC and not only in the new name: for administrative matters Unions had three votes each, so the Union vote was by no means insignificant as compared with that of the more numerous national adherents. These voted on a one country – one vote system and remained, as with IAA and IRC, the backbone of the organisation and the main financial contributors to its work.

We can look at the 'Purposes' which opened the Statutes here and compare them with the declared purposes of some Unions. The ICSU purposes were:

(i)   To co-ordinate the national adhering organisations, and also the various international Unions;

(ii) To direct international scientific activity in subjects which do not fall within the purview of any existing international associations;

(iii) To enter, through the national adhering organisations, into relations with the governments of the countries adhering to the Council in order to promote scientific investigation in these countries.[15]

Purpose (i) is the weakest in tone and the strongest in real meaning. The world need was for the bringing together of scientific communities by some centralised effort, not reliant on the whim or good will of individual academic bodies or charities. Purpose (ii) seems at first to revive the intention of IRC to initiate research, but IRC had wanted to add to existing established work. ICSU wanted to generate new kinds of work which could then be left to new Unions that were designed for it, or to collaborative efforts of existing Unions. This seems to mean that ICSU would take on the responsibility for managing the new work. In the event, still a long way into the future, there would develop a variety of interdisciplinary functions which did not yet play a part in ICSU thinking. Purpose (iii) saw the need to generate scientific activity within countries which had shown a willingness to be part of a world scientific community.

The Statutes recognised, implicitly, that the world was changing in many ways, one of which was the growing independence of territories which had come under (mainly) European rule during the territorial expansions of the 19th century. 'Independent scientific activity' was to be the criterion for separate membership, 'independent', that is to say, of the supervision of a scientific establishment of any other sovereign entity. The

British 'Empire' offered the best example of the growth of such independence, but it was not alone, and the draftsmen of the ICSU Statutes foresaw, it seems, other such changes.

These ICSU Statutes did not impose any pattern of Statutes on the Unions, although Union Statutes had to be acceptable to ICSU. They varied widely. For example, IAU worded its purpose as simply to facilitate relations between astronomers and to promote the study of astronomy in all its departments. IUGG referred simply to co-operation in the study of the shape of the earth, and then to the comparison of instruments used in different countries. IUBS included reference to laboratories, and to the organisation of international excursions. The short-lived IMU started its purposes with the high-minded 'the encouragement of pure science', but added more down-to-earth considerations.

Since the ICSU General Assembly and the Executive Committee were now to have significant Union representation, we can gather from this variety of Union purposes that there was going to be no dull uniformity of ambition in the Council. This seemed to meet the aspirations of those who, like Lorentz, had seen science as the real force in this area of international action rather than the policies of governments.

Independence showed itself in another way. There had been a prolonged effort on the part of the League of Nations International Institute for Intellectual Co-operation (IIIC) (the former Committee) to draw IRC into its orbit, but this had been resisted by IRC, courteously but carefully. ICSU now adopted the same attitude. The League of Nations' efforts to develop international cultural links were less impressive to the scientists than their own, however praiseworthy they may have been in other eyes. IIIC was not able by its constitution to carry out the same kind of practical work in a limited field as, for example, IUPAC. It had a predominantly literary and philosophical outlook. Its proposals and meetings were aimed at such things as library co-operation, which, although useful as service elements to the scientist, were not directed at generating scientific activity.[16]

An ICSU Committee was established to consider ways of co-operating with IIIC, and a form of agreement was proposed entailing such elements as ICSU having the status of a consultative organ of IIIC: ICSU to consult on manifestly scientific matters as it saw fit; each to have right of attendance of an observer at each other's General Assemblies. An agreement was eventually signed, but soon became inoperative because of the outbreak of war and the eventual demise of the League of Nations.

Finance

The financing of ICSU had strength and weakness. The strength lay in its domicile being in Brussels and therefore under the Belgian law which allowed to international bodies some freedom from tax constraints. One weakness was that which plagues any Treasurer of an organisation dependent on subscription income from members who are themselves dependent on other subscription income. Another weakness lay in the unreliability of contributions. The world's money had become less and less stable in the years after 1918. In the 1920s movements in exchange rates and the alteration of standards were more frequent than the sluggish yearly cycle on which organisation treasurers had to function, and they cannot be blamed then (or even now) for seeking some sort of anchor for their accounts. Subscriptions are not only a factor in the functioning of ICSU and of the Unions, but are an index of their general progress.

IRC members had subscription obligations and concomitant voting rights dependent on populations, which, if they included dependent territories, could be calculated so as to be out of proportion to the size of the effective scientific community. (Belgium and its Central African possessions, for example.) ICSU went for a one country–one vote system which lasted until 1946.

By the time of the General Assembly of 1931 everyone had seen the effects of a financial turmoil unequalled for a very long time. The unit of subscription was therefore set in gold: the annual subscription for each country was to be 100 gold francs. It is not possible to give any clear idea of an equivalent purchasing power in present-day terms, but this gold basis was at least something which, seemingly, would be affected only by larger economic movements, and not by mere money-market surges.

The income of each Union was originally based on the population of each member country, but this had been modified in most cases by successive General Assemblies of the Unions. Each Union was therefore to subscribe up to 1.5% of its annual revenue, the actual proportion being reviewed from time to time. It all seems very simple, but it never was. Non-payment, by some Unions and by some countries, was already reported to the 1934 General Assembly. This red thread runs right through our historical weave. At first, the contributions of the national members amounted to the bulk of the annual income, the remainder being provided by the Unions (but see Appendix 2).

Subscriptions carry voting rights. The ICSU Statutes were simple on the surface: one vote to each country, three votes to each Union, on matters not of a purely scientific nature. On scientific matters, it was to be one member–one vote. (If difficulties were to arise it would be on agreement whether a particular question under debate was or was not purely scientific.)

Membership

In the first instance only national and Union membership was envisaged. This was quite enough to digest: in any case the limitations on style of membership meant that the Council could keep a tight hold on activities which claimed to be ICSU-based. Members could be certain they were dealing with other bodies like themselves. Eventually this would change by the creation of, *inter alia*, associate status which did not entitle to a vote: there was going to be a wide variety of bodies related to ICSU, working with it, under it or beside it. The ultimate financial and administrative authority resided, of course, with ICSU, once it had been approved in principle in the General Assembly by the representatives of the national members and the Unions. ICSU was to decide how its own income should be used.

The ICSU Statutes spoke of Unions as if they were all simple bodies. None actually was. There were variations of form and function, some of substance, from the beginning, in the many sections which made up all the various Unions. The eight Unions (seven when IMU temporarily subsided) represented a much larger number of constituent bodies each with a life of its own, e.g. IUGG. For example, there were the several Associations of the IUGG, each of which had an independent origin, a largely autonomous life, and an independent substantial programme. The paper pattern of organisation by no means represented the realities of the work of the separate scientific communities indicated in a Union title. The chemists in particular were continually finding that the pure chemist and the applied chemist were drawn into useful dialogue. An association of manufacturers of fat and fat products wished to benefit from contact with their theoretician colleagues and found themselves extending their own work into the theoretical domain. An area which was bound to expand was that of the many substances of pharmaceutical interest, natural or synthetic, so IUPAC found itself establishing relations with the International Pharmaceutical Federation outside the ICSU family.

However, there was no argument about qualification for Union membership in these early stages. The Unions which existed had established their own character, and seemed to be somewhat alike in function.This did not change very much, so that ICSU remained in the hands of a relatively small number of Unions which understood each other's character, in teamwork with the national members.

Since 1919, national membership of IRC had changed in numbers more than Union membership over the years, but the balance began to change even more when the consequences of war changed a great deal of the world political and intellectual pattern. As well as mere numbers there were other changes, such that it was not always the same kind of governments that sustained the academies.

There was one gap of painful significance. Germany, which had contributed so much to science, and had played a leading part at the turn of the century in laying the foundation for international action, was still not a national member. At the 1934 General Assembly, however, there were three German invited guests, two from the Prussian Academy of Sciences, one from Berlin-Dahlem. Austria, too, remained outside. This was sad. The country was no longer the Austria of the events of 1898, but its Academy of Sciences was in the true line of succession.

The scope of ICSU was bound to be extended, but it happened in several ways. One was the internal diversification of Unions, as each area of science grew through new discoveries. Another was the eventual founding of new Unions, but there were to be no new initiatives for some time to come. Up to 1945 the Unions (with one important exception) represented a somewhat traditional partition of science into familiar fields which satisfied most needs. The exception was URSI, which had been created in 1910, when a new science connected with electromagnetic radiation gave rise to a new technology of communication. There were to be more examples of this kind of relationship in the future. A third means of extension was the change in balance of national interests, a factor which was to be important after the 1939 war, when environmental interests and certain human interests (exemplified in new Unions and in IBP) gradually came into prominence. The geologists still remained content with the service provided by their strong tradition of quadrennial International Congresses.

Choice of legal domicile

Brussels had been a good choice for the domicile of ICSU for financial and tax (and possibly political) reasons. Its value evaporated temporarily with the 1940 occupation, but the suspension of normal scientific activity masked the need to reconsider this question because so many others had to be answered before long. With the resumption of normal communication in 1945 it seemed that a Brussels domicile was once again desirable and it remained so, even though the actual administration moved successively to Paris, London, the Hague, Rome, and then subsequently to Paris. Where the actual administration was carried out up to 1939 and through the war years was a different matter.

Legal domicile and place of work are by no means the same thing. The style and manner of the day-to-day work of administration of ICSU have changed from the very personal to the efficient teamwork of the present day, but at this point one need only say that in its early days the office of ICSU (like that of IRC) went round with the Secretary-General, provided by him out of personal resources or out of the staff support provided by his employing or academic organisation. This remains true for most of the Unions to the present day. Six of them, such as URSI and IUPAC, can maintain permanent administrative offices, but in most cases, administration has to be carried out by the Secretary-General for the time being. This may influence effectiveness and can sometimes, regrettably, be a bar to the election of a worthwhile candidate who is unable to call on local logistic support.

Conclusion

So much for the administrative skeleton of ICSU. The operation of transfer of authority from IRC to ICSU worked very smoothly, perhaps because IRC ought to have had the ICSU form from the beginning. This is easy to say with hind-sight, but one has to remember that IRC had itself promoted change, from a caucus of academies to a nation-based international system. The ICSU structure went further in doing away with exclusivity and in recognising the autonomy of the Unions. Before it too was subjected to a review of form and function in its turn, it had a decade and a half of eventful scientific life to live. Let us turn to that scientific substance.

Chapter 4
ICSU development: 1931–1939

The history of ICSU is not simply a footnote to the history of science, but, to change the image, a counterpoint. The remarks that follow refer mainly to the physical sciences, but the same exercise could be carried out in biology or technology.

Take any decade from 1500 to today and, if you pick out the right topics, you can make it look like one of the most important in the history of science. It is hard, however, to deny to the period 1931–40 the merit of absorbing interest. Discoveries were made in those few years that generated work which is carried out now on a world-wide scale, and has stimulated the formation of new international groups of the kind for which ICSU is the centre.

We remarked that the history of ICSU is full of people, who have to meet each other either face to face or by writing or speaking or, in recent times, using electronic apparatus. About the time that IRC was turning into ICSU, some shared influences were felt in many areas, not just the scientific: colleagues in many callings found it easier to communicate, and, having enlarged their horizons, wished to consolidate their connections. The congress remained one of the most valuable formats, but continuity and the support of related smaller groups became more and more important.

A difference was being felt, however, in verbal communication. Immense numbers of words moved constantly on the international wires and submarine cables, conveying messages, data, and information. The telephone had spread to some point at least in every settled part of the world. If you wanted to reach a colleague who was near a telegraph office or a telephone you could make contact and have a response, even limited in content, the same day.

You could arrange to meet a colleague in person, with fair speed, if you had a passport. The ease with which one could cross a frontier had diminished in some respects. There had been a time when movement across frontiers was virtually unrestricted except in time of war, but now passports were needed and a visa was becoming a necessity for some entries. The exit visa hindered some who wanted to move out of their countries. There were, after 1918, a great many people who could not easily identify themselves as nationals of any country, and there were many efforts, including some by the League of Nations, to devise means of dealing with these dispossessed. There were at first two problems on different scales: that of mass displacement as in Russia, and personal displacement. The mass displacement is beyond our scope. By the 1930s the problems in the scientific world were mainly those of individuals who could be identified as belonging to the scientific community. Political changes in Europe brought many refugees into Western European countries and the United States, but ICSU had at that time no special arrangement for trying to deal with them.

By and large, movement was easy enough for large congresses to meet when enough notice was given. Some even reached a size which threatened to stretch the accommodation needed beyond the capacity of host cities.

One other change, which applied to life at large, affected the sense of participation in world events, to an extent insufficiently appreciated. This is the increase in the ease of printing photographs. The invention of the half-tone system of printing photographs in letterpress machines had striking effects on the character of newspapers, but it also meant greater use of photographs in text-books and technical journals. The little that a photograph added to a piece of news, or a description of an apparatus, made a disproportionate difference to its power to enlighten (or mislead).

A social innovation extended the general sense of participation in events, namely the cinema which provided a sense of being in touch with an outside world when it became general. The accompaniment of the moving image by sound also had a great effect. The most successful methods relied on the photo-electric effect; here is an example of a phenomenon which continued to be the subject of intense theoretical and experimental study becoming an everyday utility. Or, to put it another way, the photocell became as important to the cinema as the camera.

The scientific community's own methods of communicating the results of study and research did not change very much in style. Learned journals were slow to adapt to new techniques. However, the change in the apparent size of the known world, which gave everyone in the literate world a heightened sense of being 'in touch' also affected the scientist by imparting a private sense of being increasingly in touch with one's own kind.

New international bodies

ICSU was not remarkable in appearing on the cultural scene in the decades between the wars. At a rough count there were more than a hundred substantial official international bodies in existence at the time of the founding of the League of Nations, some of which already had a long history of development behind them. One that perhaps touched more people than any other was the International Postal Union.[1] There were many more unofficial international bodies. A sense of community was an important factor in their development in the period we are looking at.

The International Olympic Committee, founded in 1894, was outstanding in its development in the interwar years into a truly world authority, with a continuity and scope that many another organisation must admire and envy. (Its importance became clear when it was used for political purposes in 1936.[2]) The lawyers developed the International Bar Association. The International Amateur Athletic Association which had been founded in Stockholm in 1912, combined in 1936 with the Fédération Sportive Féminine, founded in 1921. The International Society for Contemporary Music (ISCM) was founded by a group of enthusiasts in 1922 and quickly became influential.

In the cultural field the International Federation for Documentation (FID) was established in 1938, but had its roots in the Institut International de Bibliographie which was created in 1895 by two Belgian lawyers (Paul Otlet and Henri Lafontaine) to promote systematic documentation methods. In 1905 they began to promote the Universal Decimal Classification (inspired by the Dewey system for library classification), which has a world-wide use, still being developed by the FID from its headquarters (now at The Hague) but now as only one element in a large programme of research and development in the techniques of printed communication. One of the tasks taken on by the League of Nations Institut International pour la Co-opération Intellectuelle (IIIC) was the encouragement of links between museums. It issued a few publications from 1922 to 1940 including a periodical MUSEION. (After the formation of UNESCO, the International Council of Museums, ICOM, was created to continue and develop work in this field.[3])

The International Labour Organisation (ILO) was created in 1919 under Article XIII of the Treaty of Versailles (and was to be the first agency to be affiliated to the United Nations, in 1946). An International Bureau of Weights and Measures, which had been set up in Paris in 1875 was modified in constitution in 1921, and ran a successful and authoritative series of conferences every four years.

International action in the technical sphere can be seen in several areas. The International Telegraph Union of 1865 had established standards of method and use of the world-wide telegraph system. By 1932 radio and other technical advances prompted the setting of up a new International Telecommunications Convention (effective 1934) which combined and superseded existing International Telegraph and International Telecommunications Conventions.[4]

An international meteorological organisation would seem to be a real need, and there had been many congresses throughout the 19th century. Regular meetings of heads and staff of national meteorological bodies led to the adoption of world meteorological conventions through the International Meteorological Organisation (IMO) which was founded in 1873. (This provided one of the bases for the establishment of the World Meteorological Organisation (WMO), given specialised agency status by the UN in 1951.) World-wide activity in seismology dates back to the 19th century, and on the initiative of J J Milne there had grown up an International Seismological Summary, providing a continuing record of observations carried out in many countries.[5]

Outstanding in its contribution to the spirit of international co-operation was the Second International Polar Year (1932–3). (The first had been in 1882–3.[6]) Its memory and influence survived and led to the inception of the International Geophysical Year (to which we pay attention in a later Chapter).

So much for the general climate of internationalism. We shall deal with some of these events and organisations later. It has been necessary to take some of our comments beyond the end of the 1939–45 war for the sake of perspective. The war brought many things to an end, but, as will be realised from this sample of organisations, there was also much continuity.

The first eight years of ICSU

IRC had a useful life of 12 years. ICSU had only 8 before it had to hold its breath and then start again. However, what it did in those 8 years made it possible for it to survive and revive.

Looking back from the last decade of the century it is sometimes difficult to get the flavour of daily life as lived in the decade between the formation of ICSU and its hibernation in the intellectual winter of war. In the academic scientific world the ordinary business of teaching students, supervising theses, and searching for research funds went on in the 1930s, as it had done for a long time past, or so it seemed.

In the world of applied science there was more change apparent. The world financial structure had been shaken, by the United States Stock Market crash of 1929 and other economic upheavals, the origins of which are still disputed, but may well have lain in the errors of judgement of those responsible for the post-war settlements. It is not possible to separate all the many factors in the pattern of industrial development.[7] In many countries there were gradual changes taking place in the way science was put to industrial use. Some large companies, particularly in the old but rapidly expanding chemical industry and the new field of electrotechnology, were developing large research organisations. In some countries experiments were taking place in the creation of new kinds of research organisation independent of both universities and industries. The organisation of applied research was even developing its own literature and philosophers.[8]

One aftermath of the 1914–18 war had been the persistence of old enmities which weakened the development of ICSU's European relations. On the other hand the rising tide of United States science was strengthening science at large, more and more centres of excellence becoming apparent, and more and more publications finding willing producers and outlets there. The scope of many of the Unions was substantially widened by an increasing volume, in many countries, of industry-based research as well as of academic research.

Inventiveness was as vigorous as ever, and life was being changed in style for many people even if the level of personal prosperity was miserable for many and insecure for almost all. In the advanced countries, movement and communication were, perhaps, the greatest changes to be visible on the surface: private motoring, commercial airlines, mechanisation of agriculture, radio, all contributing to the modification of common life by technology. In the less advanced countries awareness of differences was deepening, and there were to be political changes the speed of which was to be catalysed by the Second World War.

In the economic field, the impact of technical development was being felt more and more. A single contrast may serve for many. The impetus given to aircraft development by the 1914–18 war continued, but equally important to the ordinary person and to the average industrialist was the growing availability of small electric motors, which super-seded mechanical transmission in factories and made possible the invention of machines like the domestic vacuum cleaner. Few people flew, but millions of house-wives blessed the name of Hoover. One could multiply instances of the spread of inven-tion on all scales.[9]

Radio, which had been little more than experimental in the year IRC was born, had become an important means of information by the time IRC turned into ICSU. If one wants to point to a single thing showing how applied science had moved into people's lives, consider the broadcast time signal. Into this simple, brief daily event went the skill of the astronomers, the electrical engineers, and, of course, the skill of international negotiators.

These are things which touched ordinary daily lives. Remote from them, but of fundamental importance, was such highly specialised work as the preparation of the *Carte du Ciel*. This mapping of the heavens, intended to be complete, had been going on since 1887, and it was reported at each of the General Assemblies of ICSU that it was that bit nearer completion of execution but still lacked quite the resources for com-plete publication, an almost asymptotic approach to a good end.

ICSU embraced, on the one hand, activities long known and well understood, and, on the other hand, activities generated by new discoveries, applications, and ambitions. Much of its earlier scope was naturally in what one might call the older generation of science. Striking discoveries were indeed made in the 1930s,[10] but the accent of inter-national co-ordination of work in them did not fall until after the caesura of the 1939 war, and the wide extension of work in such fields as atomic physics, genetics, or radio-astronomy.

There were at least four recognisable forces at work: the intellectual desire to make international scientific links stronger, the political impulse to promote international action on an official level, a desire to achieve political ends by economic pressure, and the choice of military force. As IRC was turning into ICSU the world political pattern was changing in a manner which was bound to affect every such body. The economic contraction affected traditional large-scale extraction and manufacturing industries, at the same time as new science-based industries developed, although without the capac-ity to absorb the growing numbers of unemployed. The flux of the political and diplo-matic background is familiar: the rise of dictatorship, overt or concealed, the failure of the League of Nations to be effective militarily at the same time as it increased its authority in social issues, the modification of the European character of the League by the cautious acceptance of the USSR and the resignation of Germany. The life and fate of the League of Nations was to be important to ICSU because the discussion of its successes and failures, strengths and weaknesses, and the form of its organisation were all to play a part in the post-war examination of new bodies. The self-imposed responsibilities of ICSU can be seen, then, in such a setting. We can compare the expression of hopes and achievements at each of the successive General Assemblies of the Unions and of ICSU itself from 1931 to the latest date, before war broke out, at which each of them was able to hold such a meeting. The last task of most General Assemblies is to propose a time and place for the next one. Some of the proposals were for dates which turned out to be in time of war.

The Unions in the 1930s

IAU

The International Astronomical Union can be excused for beginning its reports[11] with expressions of pride in its ancient forebears, but it can also be praised for the dogged perseverance of the astronomers who had a near genius for compiling minutely detailed records, of which the *Carte du Ciel* was one of the most important products. One of the older tasks was that of settling an agreed Lunar Nomenclature, and this came to an end in this period. This had long been an enthusiasm of a few, but it was a necessary task which could be wound up. An endless need was that for the redetermination of longitude, and for this new means were being tried out. The second quarter of the century saw, for example, the transmission of time signals by telegraph replaced by transmission by radio, this being a new resource for the determination of longitude. The quartz clock was also being received with enthusiasm.[12]

Other techniques were coming into use. For example the spectrophotometer, which had been known since the late 19th century, was now taking on a new role. The devising of useful forms of the instrument encouraged the use of standardised procedures in numerous observatories, allowing closer comparison of results than had been possible hitherto. This field gradually grew to the point that it was feasible to hold a symposium devoted exclusively to it as part of the 1935 IAU General Assembly. Out of this came agreement on standardisation of certain optical filters for spectrophotometric use. This is just the sort of widespread co-ordination which, although not unique, could now be achieved quickly with the Union type of organisation.

Although the old political breaches remained, the astronomers were forced, by the nature of their science (its need for collaborative observations over long distance), to set an example in restoring contacts.[13] The Catalogue of the Astronomische Gesellschaft was still being produced in Prussia, and some observatories (including some in Russia) began to send data to it. Minor Planet observations were co-ordinated by the Astronomisches Recheninstitut in Berlin which had a link with a 'Students' Observatory' in Berkeley.[14]

A special source of strength to the astronomers was the benefit from the work of amateurs. The IAU kept up good contact with amateurs in many places in the world. On the professional level the astronomers also worked with the Geodetic Association of IUGG in observing the motion of the earth's axis.

The astronomers also experienced their surprises. Optical astronomy had reached a high degree of sophistication with the phenomena of refraction and reflection, of photography, of spectroscopy and spectrophotometry all extensively exploited. Then in 1932 a chance observation by Karl Jansky that some radio interference could not be attributed to terrestrial sources led (although not immediately) to the development of radio astronomy, and to the investigation of the generation and sources of this part of the radio-frequency spectrum outside the earth.

The astronomers and astrophysicists benefited from the new theories of the atomic nucleus. The generation of the heat of the sun and the cycles of stellar birth and decline could now be related to ideas about the other end of the scale of entity.

IUGG

IUGG's several sections had long experience, the Geodetic Association going back to its foundation in 1864 by the Prussian, General Baeyer.[15] The figure of the earth is a classical problem, offering elegant opportunities for relating geometry to experiment in several forms. The surveys needed to relate the mensuration of the earth's surface to the geometry of theoretical figures for the earth (the geoid) are complex and need to be carried out on an immense scale, even continent-wide. By 1937 the geodesists were facing up to some interesting diplomatic problems, because their desire to extend measurements of longer and longer arcs of meridian brought them up against frontiers of countries which did not easily respond to the call for scientific detachment. The Chinese government of the day offered to co-operate in providing for measurements in the large spaces of the interior, but the condition of China at that time promised little long-term activity. The problem even became intercontinental as efforts were made to unite measurements from Canada across United States territory in Alaska, and on into the USSR.

The interpretation of the local configuration has to take account of local gravitational characteristics, and in this the geodesists found new apparatus becoming available that were transportable with relative ease, and thus allowed for comparative studies over a wide area. New principles were being introduced into many kinds of measurement and IUGG did valuable work in making them widely known.

For the layman seismology calls up images of the dramatic and tragic, but to the professional it is a matter of continuous monitoring of the movement, great and small, of an earth which is never really still. The correlation of data on simultaneous events is of the greatest importance, and in the 1930s seismologists, like the astronomers, were benefiting from the rapid transmitting of information and data first by faster and faster telegraph and then by radio. General Gustave Ferrié had been notably active in this, taking advantage of the new aerials of the Eiffel Tower. Later a radio centre was established in Strasbourg.

There was a limit to what could be done by individual observers or stations. The pioneer work of J J Milne was amplified by the Union's encouragement of national seismic services, which developed in the United States, Japan, Tunisia, and elsewhere. Efforts were made to set up seismic observatories in places (especially island sites) known to be vulnerable to seismic and volcanic disturbance. Later a centre was created in Poland, filling a prominent gap in the coverage of the European continent.

The Association of Meteorology, which formed part of the International Union of

Geodesy and Geophysics, had been founded in 1922 and had fostered international codes for transmission of information about meterological conditions, and proposed to review scientific methods of weather forecasting as applied in different countries. Meteorology was a subject of which the intelligibility appeared threatened by multiplicity of data, so in 1932 the Association decided to bring some order into meteorological discussion by advance printing of a limited range of papers for its conferences. It also promoted standardisation of methods (for example in rainfall measurement) and of instruments used in different countries. It co-operated with the International Meteorological Organisation (the forerunner of the WMO) in many ways, for example by granting subventions for work on the International Polar Year (1932–3).

In 1937 we hear the first mention of two subjects, which were to become of great importance. One arose out of the far-sighted realisation that commercial trans-Atlantic flight was going to be extensive and important. A call was made for improved observational facilities, sea-based and land-based in Iceland and the Azores. The other was air-pollution, but the principal pollutant they then had in mind was in the form of dust, volcanic in its most striking manifestations but also coming from other natural sources and from industrial activity. The present-day concern for complex chemical reactions in the atmosphere had not yet set in.

The International Association for Terrestrial Magnetism and Electricity was another IUGG body that had its roots in the Brussels meetings of 1919, but it had only taken on definitive form with the strengthening of national representation. It had its own links with the informal International Meteorological Association, but it sought also to work with the Meteorological Section of IUGG, with URSI, and with IAU. It was much encouraged by the success of its work in the Second International Polar Year, especially as regards standardisation of magnetic instruments. In this it could claim important status because of the extent of magnetic phenomena and the importance of magnetic mapping in navigation. As regards the upper atmosphere there was much to be proud of, the ionised layers (Kennelly-Heaviside and others) becoming better and better characterised and understood. Cosmic radiation was another topic receiving increased attention, with the proofs that the source was extra-terrestrial and that it probably consisted of charged particles.

The International Association of Physical Oceanography settled on this title only after some discussion of alternatives. Founded in 1919 by the enthusiastic Prince of Monaco (who died in 1922) it eventually established a programme of collaboration between many bodies, concerned with particular seas (the Mediterranean, the Atlantic, the Pacific), or particular limited problems of wide significance such as the determination of mean sea-level, standardisation of chemical analysis methods, or the study of tides. Its greatest strength was in publication, of which it eventually offered a wide range.

The Association of Vulcanology (so spelt at that time, nowadays Volcanology) founded in 1919, also urged, from 1931 onwards, the importance of collecting and disseminating information, and was one of the only organisations to note the value of a museum of specimens. It encouraged the collection of information on active volcanoes, the substances emitted, the temperature, and other characteristics of lava, of fumaroles and hot springs. It urged the gravimetric study of surrounding terrains. It also drew attention to the importance of the study of atmospheric dust. Requests were made to many countries that ships of their merchant and naval marines should report submarine eruptions that they observed. Only Great Britain, then the largest marine power, appears to have made any active response to this.

The International Association for Hydrology had as its scope one of the fundamental needs of mankind, namely water supply and the relation of water to the habitable and unhabitable terrain. It recognised its own value in three areas: the production of motive power, irrigation, and in warning of floods. The study of rivers took first place in its interest. As in so many other sciences, the study of effective and reproducible instruments was a constant preoccupation. The movement and temperature of lake waters prompted new instrumental studies. Water in the form of snow and ice had equal interest, and the Association had close relations with the glaciologists. It hardly needs saying that the Hydrologists also found much common ground with the meteorologists.

In 1936 a new Committee was inaugurated: Committee on Continental and Oceanic Structures. This foreshadowed the growth of interest in major structural problems and the origin of the continents which was to be a feature of post-war geology.

IUPAC

The Union representing chemistry differed from all the others in one important respect: industrial chemists played a part in discussion and decision all the way from the beginning of the century.[16] In 1918 it had been proposed to convert the existing association of chemists into an International Confederation of Chemists, but the formation of IRC had inspired the formation instead of a Union within the IRC structure. The importance of applied chemistry was recognised in the title adopted: International Union of Pure and Applied Chemistry. (It is a sign of the times that the physicists followed suit: to form the International Union of Pure and Applied Physics.) However, in 1928 the name was changed to International Union of Chemistry, suggesting that the industrial chemists no longer had any considerable voice in its affairs. (This was reversed in 1947, the name going back to IUPAC.)

Germany had joined the Union in 1931, as had the USSR, and the Union had a German Vice-President at its General Assembly in Madrid in 1934. The welcome was given by the President of what was, at the time, the Spanish Republic. Membership fluctuated: Australia and South Africa had resigned, Estonia and Lithuania resigned and then rejoined when IUPAC devised a new system of dues that put these two countries in a category they could afford. The 1934 meeting also spent much time on the question of photocopying, and worried more about the law of copyright in various countries than it did about the practicality and economics of available processes.[17]

At the Rome meeting in 1938 we see the beginning of a new approach to science: the chemists agreed that they should consider the social aspects of their calling, and the consequences for the general public of the practice of chemistry by the specialists. The chemists were to be at the centre of the great 'science and society' debate which lay in the not very distant future.

The scientific discussions show the development of chemistry during these years. A Commission on Atomic Weights might have been expected to carry on a well-established tradition, but now we find a Commission on Atoms, the difference being that this new body was concerned with those aspects of atomic character which had emerged since the beginning of the century: isotopes, and the dimensions of atoms. The nomenclature and terminology of chemistry was also brought under close review, as were the many constants which formed the basis of chemical calculation.

The atomic theories had their effect on chemistry. The Bohr-Rutherford atom had offered chemists an image of bond formation which provided a basis for generalising structural chemistry, and also suggested mechanisms for reaction. Not that there was any general acceptance of any one interpretation of new theory. The chemical bond threw up as much controversy as any theory in politics or religion, but the overall effect was one of progress from the merely descriptive to the usefully predictive. The conventional methods of chemistry were also extending their scope, with the determination of the obscure composition of many substances of natural origin.

There was also a development in the concept of the chemical molecule itself. Much of chemistry had grown out of the study of small groupings of atoms, but during the 1920s studies were made of very large molecules, first by the study of known substances, and then, through new synthetic methods, by the creation of large molecules of novel kinds. Some results were eventually of world-wide influence. Plastics of several kinds had been known since the late 19th century, mostly cellulose derivatives, their properties being due to the large molecules of natural substances. The first to be derived from non-natural sources was the phenol-formaldehyde condensation product later known as Bakelite.[18] This is an example of more than one worker arriving at the same conclusion or the same process at the same time, the Englishman Swinburne filing his patent in 1907 only one day later than the Dutchman Leo Baekeland (after whom the commercial product was eventually named). This inaugurated an era of novel pathways to new materials (and, incidentally, a new era of electrical insulating materials). Innovation relied on rather hit-and-miss methods, but several industrial firms sought methods of controlling polymerisation processes with the aim of producing substances of predictable properties. The most brilliant success was that of Wallace Hume Carothers in the Dupont laboratories, producing the first polyamide, which became known as nylon.[19] Commercial exploitation of the material was delayed until 1940, but the science of the successful process established the pattern of the creation of a new family of synthetic materials of manufacture and fabrication.

Another synthetic material was discovered, partly by accident partly by design, in the same period. A study of a residue produced a synthetic inert substance of great electrical interest, namely polyethylene, again the first of a family of substances deriving from the new ideas of molecular structure which were emerging. Polythene was first seen experimentally in 1933; it was being manufactured in 1939, and soon became of great importance in electrotechnology.

There are other examples of widespread interest in new kinds of investigation, but they did not all have the same style or objective. For example, the search had been going on since the beginning of the century for drugs which would be effective and specific against bacteria. In 1932 Domagk had successfully examined a new dye, prontosil, for anti-bacterial action withholding publication of a careful clinical study for some time. Then workers in the Pasteur Institute in Paris had shown that the effective agent was not the prontosil itself but a breakdown product, sulphanilamide. Many companies set to work to benefit from this discovery, success being achieved by members of staff at May and Baker, yielding sulphapyridine, M & B 693, effective against pneumonia, the great killer of the normally healthy as well as of the weak, both young, and old.

Let us generalise for a moment, taking this topic of medical chemistry as only one example of a general issue. In 1928 occurred Fleming's first observation of penicillin, its first confident clinical application not taking place until 1943. There has been a great deal of discussion of the apparent delay in the recognition of the value of this splendid life-saver, which only entered its maturity with the work of Florey and Chain[20] 15 years later. Why so long? A response to this question would take another book, but it is certain that the need for an effective bactericide for use in wartime gave an important impetus to the development of penicillin into a usable form. But we can ask the same question about Mendel and the delay in the recognition of his genetic systematics. Cast about further and we recall Avogadro's hypothesis devised in 1811 but only welded into the system of chemical theory after 1860.

If we have spent rather a long time in discussing the progress of chemistry it is because the author was once a chemist; someone with another initial training would find quite as much to say in his own subject, and probably would come to the same conclusion: the development of science does not follow any regular, logical sequence of events. Science does not march. It progresses by hop-skip-and-jump, by hesitation and confident stride, chemistry as much as any other science. So does the history of ICSU or any of its Unions.

URSI

We have remarked on the rapid growth of the science and techniques of radio, the outgrowth of the highly theoretical studies of electromagnetic radiation of the later half of the 19th century. More than in any other field of scientific activity radio science was international, indeed supranational since its emanations recognised no man-made frontiers. We leave aside here the public and commercial uses of radio: URSI was concerned with radio as a scientific study only, and a central problem which prompted much ingenuity in its solutions was that of the generation, maintenance, and transmission of definite frequencies. The modern period begins with the invention of the electronic triode oscillator. The improvement in the measurement and establishment of frequencies is related to the measurement of intervals of time, so the radio scientists found themselves closely involved with the astronomers and geodesists. An example of the interaction of apparently unrelated sciences is that of the comparison of radio frequency time-interval determination with intervals determined by a pendulum used in gravity determination. Discrepancies due to a tidal effect of the moon were detected in this way.

A notable service developed by URSI was the URSIgram. This originated in the action of the French and American National Committees (led by Gustav Ferrié and A E Kennelly) in issuing plain-language daily bulletins of geophysical and solar date from the Eiffel Tower in 1923. Starting in August 1930, compressed coded bulletins were transmitted from high-power stations of the US Navy. Later other countries took part.

There was also important radio work in connection with the Second International Polar Year, but we refer to that in chapter 12.

IUPAP

The main event of the whole 1931–9 period for IUPAP was the 1934 General Assembly in London and Cambridge. It was devoted to two main subjects, each of them destined to become dominant in both pure and applied fields. One was nuclear physics, which was to take the centre of the political as well as the scientific field. The other was solid-state physics.

The 1930s period has often been described as the golden age of physics. One could apply to it the words the poet Wordsworth used of the French Revolution: 'Bliss was it in that dawn to be alive, but to be young was very heaven' (1805). Theories of the atomic nucleus moved from the detailing of the consequences of the relatively simple (by our standards) Bohr–Rutherford atom to theories of nuclear structure implying the possibility of induced change. This was the output of no one person or school. The work that contributed to the change was carried out in many countries.

Some of the names that turn up are familiar now only to those who know something of physics, although the Geiger counter turns up in popular television often enough to be both a name and a tool. Less familiar would be Anderson and Millikan's discovery of a positron during their study of cosmic rays, or Chadwick's identification of the neutron (1932). The popular press made something of heavy water (1932) when Urey isolated enough to determine its properties, but its significance dropped from view until much later when its role in the generation of atomic energy was widely canvassed. To the physicist and chemist, however, heavy hydrogen, deuterium, was, from its first naming, a substance of the greatest importance in practice and theory.

The physics of the solid state[21] was to transform much of chemistry and biology, and, in the study of electronic structure of crystals, was to transform applied electronics. This profoundly influenced communication on all scales and in all media, as well as penetrating into all aspects of scientific method, its theories and its instrumentation.

With the extension of atomic theory went extension of experimental resources: for example, the van de Graaff high voltage generator (1931), Lawrence's cyclotron (1932), the particle accelerator of Cockcroft and Walton (1932), were beginning to grow in size and power, but some of the subtlest experiments were still carried out in apparatus small enough to be accommodated on an ordinary laboratory bench. Perhaps the smallest device was the photographic emulsion itself within which particle tracks could be observed, analogously to observation within a cloud-chamber. There gradually emerged through the work of Joliot and Joliot-Curie, of Hahn and Strassman and others, the possibility of that extreme intrusion into the nucleus which was alluded to above. Atomic energy then became for a time a subject of which there could be little open discussion.

IGU

From the 1934 report to the International Geographical Union by its Secretary-General (Emmanuel de Martonne)[22] we can pick out a few words which illuminate the whole history of modern science. 'Long considered as a descriptive science, geography has, over the century, become an explanatory science.' This applies not only to geography, but to the whole range of scientific work. The change from description to explanation characterises much of science in the century after the foundation of the specialist learned societies.

Martonne referred to the many geographical congresses that had been held, supported mainly by members of geographical societies whose idea of 'geography' fell short of the ideals of scientific research. Martonne may have been a bit unfair to the geographical societies, who were, after all, supporters, among other things, of many resourceful explorations,[23] but one can see how the IGU aimed to promote a scientific attitude by being the main controller of the International Geographical Congresses. It had been doing this very successfully since 1925, and had progressively established commissions for many topics, ranging from climatic variation to the geographical study of over-population.

Human geography was becoming an important issue. Martonne claimed that, even though the scope of study so far had been limited, it had been shown that, in spite of their complexity, questions of human geography were susceptible of exact study.

Another area of wide relevance to which Martonne referred was that of arid regions, and the related problems of hydrology. This, he hoped, would lead to the sharing of interest and study with geologists, meteorologists, and even economic historians. Here again Martonne anticipated a pattern of co-operative inquiry which was to become general within a decade or two.

Over the next few years, from 1931 to the war, IGU made good progress in this ambition of developing exact studies. We can pick out just one example of a technical advance of wide interest and applicability, the work of an IGU Commission on Phototopography. By 1939 this Commission had published reports on instrumentation and on the general principles of Stereophotogrammetry. Boundaries between pure and applied science were no longer to hamper this kind of work, as was shown by the importance this subject acquired in war.

IUBS

The International Union of Biological Sciences took some time to justify its comprehensive title. A Zoological section had some difficulty getting under way, but a Botanical section was vigorous and productive. It now took a lead in organising Botanical Congresses (which had originated in the mid-19th century). The Sixth Botanical Congress was held in Amsterdam (in September 1935). During this Congress IUBS held a General Assembly of its own. Botany is a subject which relies very heavily on efficient publication, and this occupied much of the attention of delegates. A paper by S J Wellensiek on 'The improvement of publishing botanical work' showed how botanical publication had proliferated in the years since 1900, and had increased by some 12% from 1930 to 1934. More than 50% were written in English, but the proportion in languages other than English, French, and German had increased slightly. He made proposals for the standardisation of style, and of dimensions of paper. He also urged the improvement of abstracting and cataloguing in the face of the impossibility for any botanist of keeping up to date by direct reading.

At the same General Assembly a short but prophetic paper was read by H Humbert[24] on 'The protection of nature from the biological point of view, in tropical and sub-tropical regions'. The principal causes of change he noted was clearing by fire for replanting. Dr Tienhoven referred to a congress on conservation in Africa called by the British Government in 1933, in which proposals had been put forward which had been accepted by many other governments. The emphasis at that congress had been on animal life, but the botanists also were encouraged to think of conservation in botanical terms because of the intimate connection between plant and animal life.

'Only connect'

There were many occasions when no one could see certain connections that we now see as logical. While some chemists were looking at large molecules created in the laboratory, others were concerned with biological structures they were not certain were molecular at all in the common chemical sense. These included some of the most important components of living matter, such as the proteins. The Swede, The Svedberg, played a leading part in this with a new method of determining molecular weights: sedimentation with the ultra-centrifuge. Some of his machines were, by the standards of the day, quite huge, operating at what were believed to be dangerously high speeds.[25] Of many findings one was significant: that proteins existed as definite molecules, very large, but not just aggregations of indeterminable composition. At the same time the techniques of X-ray crystallography were beginning to reveal the nature of other complex structures. Eventually these lines of research were to converge: we now see in the 1930s the roots of post-war molecular biology, but their character was not yet apparent.

One needs examples like this to understand some of the changes in the scientific scene which came about later. Svedberg's work had moved from the simple laboratory-bench scale into the engineering scale. It was expensive to carry out, and needed appeals for support for capital expenditure. The astronomers were familiar with the need to fund massive equipment, but here we see the same problems invading the realm of the customarily small-scale.

The range of science covered by these Unions seems limited now, but between them they did cover a very large proportion of the field of science then existing as organised academic or technical study. A recital of achievement and aspiration like this has about it the romantic touch of early enthusiasts for science,[26] but we must spoil the picture with some realism. The first years of ICSU were the years which led up to a new war, and it is nations (or would-be nations) which go to war.

National membership

Both ICSU and its Unions had a type of membership called *national* membership, national, that is to say by reason of the deceptive ease of defining membership in such terms. However, if national had meant merely territorial, there might have been no problem, but national meant also nation-state, and the behaviour of ICSU and Union members had to have regard to the diplomatic and political behaviour of these nation-states. Geography played its part, too: it was easier for a member of the Royal Society to keep in close personal touch with a member of the Académie des Sciences than with a member of the National Academy of Sciences; travel to congresses could be a slow business, so that authority fell into the hands of those who could most easily move about. (At no time has membership of academies and the opportunities for involvement been as uniform as its founders and supporters would like, but it has changed very much for the better.) All the same, scientist are citizens, and the individual members of the Council and Committees of ICSU and of the Unions had each to weigh up any personal commitment to follow national behaviour. The problems of loyalty and commitment are not confined to scientists: they are universal.

Why not a pattern of supra-national behaviour? After all, one of the consequences of the shock of the 1914–18 war had been the effort to establish international harmony though a supra-national entity, the League of Nations. We referred of earlier to the League's International Institute for Intellectual Co-operation, and to the cool reception given to it by IRC. The League itself had not been very strong or successful in its activities in the diplomatic field, but it was improving its position to some extent in the early thirties in those areas which did not involve peace-keeping confrontations. Some Unions had useful joint inquiries into terminology and nomenclature, an area in which the Commission had a wide interest. The League of Nations' IIIC was able to take part in some further discussions about joint action, but does not seem to have produced anything but agreements to have more discussions.

The present generation has grown up with a world-wide United Nations; we may need reminding that the League of Nations was not world-wide. The United States had not joined, but it had always been a leading force in the IRC and in ICSU. It also played a major part in most of the Unions. At first Germany was not allowed to join the League of Nations because of the persistence of war-time feelings. After a brief period of membership it withdrew following a radical change in style of government. The USSR was excluded at first, because of the mistrust of the new doctrines of post-revolutionary Russia's Soviet government. ICSU could therefore adopt its own independent international stance, with its own freedom to enlarge its membership. Although it did not include the most outstanding absentees from the League of Nations, except, fortunately for ICSU, the United States, the membership remained fairly stable. There were some changes, of clearly political origin. Spain had been listed as a member in 1931 and in 1934. It had gone in 1937, by which year Spain was engulfed in a civil war. The Vatican City had joined, after (but one cannot say with certainty, because of) the rupture between Volterra and the Fascist state.

However, the value of the League of Nations to ICSU was real: international action was seen to be the desirable mode of development, so that, although it might fail in the face of power conflicts, it was seen to be valid in some other important areas. The fact that the United States felt able to join the International Labour Organisation was an encouraging sign of flexibility in association.

Scientific communication

(i) Publishing

Scientific publishing has its special problems. The scientist wants any original work he does to be known and known quickly. Two related common topics came up in all Union discussions: publishing of research results and of research data. The physical limitations on printing and the financial limitations on distribution were felt by all.

This has a bearing on ICSU. One must be careful, in considering how ICSU and its Unions functioned, not to believe that all its active members were familiar with all that we can now see was going on at the time. We can identify dates and events, but we cannot be so certain about how knowledge moved about and became fruitful. One of the roles of ICSU was, and is, to promote the flux of knowledge. A closely related one was, and is, the control of the quality of published material by the mechanism of peer review.

An author of an article intended for a learned journal would generally proceed in steps, often from handwritten manuscript, or directly, to a typewritten original. From this text a typesetter using one of the machines in general use would prepare type cast in molten type-metal. This would be printed in an electrically driven machine, by direct pressure of type on paper. One must rehearse this naive account of the printing process if one is to be ready for the elaborations which were already near or well past the stage of initial invention. Most of these were adaptations to mass production of processes already known and in use for the artist producing a small edition of prints by hand, but they did not come into general use until after 1945.[27]

The printing of illustrations was now done mainly by the half-tone process, introduced in the 1890s and stimulated by its widespread adoption by newspapers.

Printing in colour by successive separate impressions of separate graphic drawings could achieve high artistic quality, but, since colour photography was difficult and fairly rare, colour photographs in the illustration of textbooks was even rarer. The problems facing organisations like learned bodies and the Unions were therefore the economic problems of using well-established systems with accepted limitations.

All the Unions had to use commercial printers and publishers for the distribution of their products. Some small-scale duplication could be carried out on crude devices useful at best for small runs of office documents and Newsletters, but learned publication had to be treated as a full-scale publishing exercise. The cost of printing and publication was a constant preoccupation of treasurers, and the mechanics of distribution and communication, of secretaries.

The amount of scientific publication increased year by year. Many studies have analysed the growth of such publication in relation to the scientific population.[28] During the pre-1939 period no easement of the publication problem for scientific work was to be found, in the absence of the kind of technical change needed.

(ii)   Transport

Actual speed of travel had not improved all that much since early in the 20th century. Records in crossing the Atlantic by sea made newspaper headlines, but by 1930 the regular voyage had not got much faster than it was in 1910. Convenience and comfort increased more than absolute speed. The same is true of rail transport. The car had not yet altered private travel as much as it was to do later, and had little effect on the ease of international assembly. Mass air-travel was still a speculative prospect, although there was some regular medium-distance travel, and some striking long-distance flights had pointed to great possibilities.[29]

Given long enough prior preparation, a very large congress could be assembled. The management of the affairs of Unions and their commissions was a different matter, because members could not often arrange to be away for the length of time needed for travel to and from a meeting. This particularly affected trans-Atlantic travel, and the involvement of the Asian and Pacific countries in work which was still dominated by the European-Atlantic nations.

(iii)   Some political considerations

Up to 1939 funding of ICSU could be said to depend largely on national budgetary provision. Contributions to Unions were very variable in reliability, so the Union contributions to ICSU were to some extent unpredictable, depending as they did on the treatment by different governments of academic staff in their employ and on their general policies towards science.

Those governments were already, by the 1930s, as different from each other as they had ever been, except in one vital respect: they now knew of each other's existence, and knew how they worked. This had not been true as recently as the mid-19th century. Japan is the outstanding example of the coming onto the world scene of a major nation. By 1931 the variety of European and American elective governmental styles was already considerable: by 1939 it was further confused by the emergence of several new varieties of autocratic rule. In Japan the older autocratic Imperial rule was being modified by efforts to adapt to western-style industry and to western-style military planning, but its academic and scientific life was making an impression on the western institutions. China was unsettled by external pressure and by internal dissent; it could not yet make any general impression on the western dominated international scientific community, although good contacts could be made, as for example in the case of geodetic measurements.

India was making active contributions to science, with its own institutions growing up, but there was still a dominant British influence. A good deal of the rest of the South East Asian and Far Eastern territories were still European dependencies with little independent scientific life, even where their geographical or biological individualities made them interesting and attractive to European and American investigators. The same might be said of much of Africa, with the exception of South Africa in which the dominant white population maintained European connections with two separate 'mother' countries, and had developed universities and other research centres of consequence. We except also the Mediterranean North Africa where French authority and influence had given a strong European flavour to much of the social activity of several countries, some of which, like Tunisia and Morocco, became ICSU members.

So in 1938–9 the pattern of ICSU's relations with its member Unions and to its national members seemed, on the surface, sound and likely to expand to the general good. Many individual scientists, however, engaged in some areas of applied science realised that events were on foot which would impose change. That ICSU survived that change is due to the fundamental soundness of the principles which had brought it into existence.

Chapter 5

ICSU in wartime

Kinds of war

The war which so many in Europe date as beginning in the autumn of 1939 was the coming to a head of an abscess which had been festering for years. Some would argue that it had been inevitable from the time of the Treaty of Versailles. Some, in what Europeans think of as the Far East, might see the outbreak of war in the Pacific as the coming to a head of a condition of war which had been turbulent for decades. In India it might be looked on as the beginning of a process of release from Imperial domination.

It is referred to as the 'Second World War' but, like the first, it began as a European war. The First had begun with a confrontation of 'Allies' with 'Central Powers'. The Second became a confrontation of a small group, France and the United Kingdom and the associated nations of the Commonwealth (the old 'Empire'), with one: Germany. Russia, which had left the old war early came soon into the new one, followed by the USA. The political pattern looked roughly the same, but there was one difference which, when it was over, was both to temper old enmities and to create new: the identification of the 'enemy' as a political system, rather than a nation-state in arms. Even when it had become a wider world war, this remained true. When the formalities had been signed, a new confrontation began, a 'Cold War'. In the East the old turbulence took on new forms.

It is necessary to say 'become'. The war started as narrowly European in territory, then spread to Russia and to North Africa, and then to Asia, the forces of other countries like the USA becoming involved as their interests and territory were threatened. Some countries remained formally neutral (notably those of South America) but could not escape the consequences in diplomacy, finance, or communication, nor in the diminution of the hitherto world-wide development of scientific collaboration. When communication was eventually restored, the appetite for scientific collaboration was greatly enhanced, as we shall see.

The origin of war is not a subject we can usefully go into here: there are too many origins and too many wars. In thinking about events in the world of science we should be wise not to forget Spain and its civil war, Abyssinia and the use of poison gas, the Rhineland, Czechoslovakia, the Austrian Anschluss, Manchuria, the Kuomintang,

Mao's Long March. While these are for other books and other discussions, they are still as important a background to our reflections as are older wars to the history of older science. There are events in the history of science which are only fully described and understood if we take war (fought or threatened) into account. Outstanding examples are the progress of nuclear physics and of rocket technology, radar, aircraft design, and computer theory.

The personal lives of many scientists were affected by war throughout the century, the enforced migration of many scientists being a notable but not unique feature. So, in our account of ICSU, the 1939 war has to be considered as a scientific war, for several reasons.

In the first place it made everyone conscious of the change that had taken place in the relation between society and technology.[1] The changes in the military machinery were enormous as compared with previous periods. Oil replaced the explosive as the crucial resource.[2] Battle swirled over wide territories, on land and at sea. (A much-quoted example is that of the battle of Midway which was fought by two naval forces hundreds of miles apart.) The deployment of the truck, the tank, and the aircraft took centre stage in tactics and strategy. Radio and electronics became essential to the cohesion of units and armies, as well as to the operation of weapons, and to the management of strategy.[3] Disease spread in patterns different from those encountered in 'normal' times and was countered by methods on which research had barely begun a few years earlier.[4] Awareness of these changes made many thinking people reflect on other changes that might take place in the conduct of a peaceful life, when it should eventually come.

In the second place, many of the controls on civil life that were imposed by the demands of war were expressed and executed in technological terms: the citizen-public was urged to think in terms of nutritional values of foodstuffs, of sources of supply of other essential commodities, of the limitations in the supply of fuel and power.

Science during the war period

The effects of the war on the world scientific community were many. In all warring nations there already existed government organisations manned by scientists who had chosen as a career the improvement of military prowess, and had chosen it voluntarily. Now the scientist was conscripted, overtly or covertly.

In conditions in which the overriding style of organisation was that aimed at winning a war, it would seem that any other style of organisation would have to take a second place and disappear. This is not so. People in every country and form of society create groups, clubs, and unions for mutual help and benefit. They can show great powers of survival, sometime over centuries, or, like the great religions, over millennia. It is not surprising that some bodies like ICSU recovered and strengthened their positions in the world at the end of hostilities.

The most important factor in the survival and recovery of the scientific community at war was paradoxical: its very involvement in the war, to an extent never hitherto imagined, prevented it from being submerged. The British already had some experience of the recognition of the significance of the scientist in the creation of the Department of Scientific and Industrial Research during the 1914 war. Eventually, after their recognition of impending war, politically minded scientists of the USA developed their own national organisations to a high degree of prominence.

The importance of war as a driving force for scientific research and technological development was most dramatically, of course, demonstrated by the production of the atomic bomb in the USA, of the V1 and V2 missiles in Germany and of penicillin in the UK. Clearly these had their effects on the structure and status of science in the post-war years. Perhaps also the seed was then planted for the popular belief that developed later that science, instead of being an unmixed blessing, was at the root of much evil.

The case of Australia is very interesting, in the light of what we have said about the emergence in the interwar years of independent scientific communities in regions which had long been subservient politically and culturally to European progenitors. The calls made on Australian science during the war as well as on its manpower and financial resources, were great. They required an expansion of the numbers of scientists, who were not likely to relapse into other kinds of occupation. This also encouraged the permanent expansion of Australian universities.[5] (To anticipate: the growth of science-based industry also contributed eventually to Australia's ability to take part in the technological adventures of space research.)

The case of the USSR is different in many ways. It is easy to represent the history of Russia in terms simply of political theory, but, however important this may be, the Russian scientific community had, certainly from the late 18th century, operated with enough constant contact with the rest of the world for its contributions to theoretical science to be part of the main stream. One only has to mention Lomonosov and Mendeleeff to bring others to mind. In applied science it suffered, as was the case in many other countries, from limitations imposed on it by socio-political and economic circumstances, but its industrial progress was substantial, and the often quoted ambition of 'the electrification of the Soviet Union' was no idle dream. The USSR had been ready, in 1939, to play a larger part in international science, by individual travel as well as by participation in international organisations, and it came out of the war strengthened by its scientists having to go through the same sort of experiences as those of other countries.

The case of Japan is different again. Its political and social history is a complex one of the changes in a class system, a military system, and a system of relations between employer and employed quite different from any met in the western world. It had been involved in wars on the Asian mainland for years, and had developed a considerable military technology, often with western advisors. It had made distinguished contributions, as we have seen, to seismology, and was gradually coming onto the world scientific scene, when it chose to involve itself in the war which had its epicentre in Europe. The end of its adventure was marked by (if not entirely produced by) the atomic bomb. It was clear to some Japanese, who had seen the war as a war of scientific prowess, that in peace it must excel in applied science, and this it proceeded to achieve.

It seems, then, that no two nations had quite the same scientific experience during the war, but all of the major nations came out of it with the scientists looking more important than they had done not many years before.

No one had woken up one morning to find a war had begun. There had been a feeling of uneasiness for years beforehand, but whatever might have been the attitude of individuals it was not part of ICSU's statutory concern to consider matters of diplomatic policy. This does not mean to say that no attempt was made to face up to non-scientific matters. On 28 April 1937, at the General Assembly in London, Professor J M Burgers,[6] speaking on behalf of the Royal Academy of Sciences of Amsterdam, introduced a resolution calling on ICSU to establish a Committee to consider 'the grave dangers that are threatening the future of our civilisation', to collect opinions and to present them 'in such a form as may help individual scientists and scientific bodies in getting a better view of the relations of their work and of their position to social problems'. On 3 May a modified Resolution was adopted as follows:

> The committee find that the appointment of a committee with the full powers suggested by the Royal Academy of Amsterdam lies outside the objects of the International Council as laid down in the Statutes. They feel, however, that within the sphere of action of the Council a committee might usefully be formed to deal with a more limited range of questions. In making the following recommendation they suggest that if it be adopted the nomination of the Committee be referred to the Executive Committee of the Council for action at its next meeting in July.
>
> The Committee accordingly recommend that a committee be formed with the following terms of reference:
> The Committee, at suitable intervals, should prepare a survey of the most important results obtained and of the directions of progress that are opening and of points of view brought forward in the physical, chemical and biological sciences with reference to:
> (1) Their interconnections and the development of the scientific picture of the world in general;
> (2) The practical application of scientific results in the life of the community.
> The work of the Committee is limited strictly to scientific activity.

The interest of this resolution lies not in any action it envisaged, but in a change of viewpoint. Its concern for the effects of practical applications lies a long way from the distaste of the Royal Society earlier in the century for involvement in discussion of applied chemistry. ICSU was accepting that it had to live in a wider world than that of the intellect. The realism of that acceptance is made clear by the resilience with which its members faced up to the growing threat of war and to its outbreak. No two members were in the same position, diplomatically, geographically, or militarily, so no common overt plan could be agreed or carried out, but all members' eyes were open, in ICSU and in the Unions, to their need for preparedness, and for survival in a time of war.

Some officers of Unions laid plans many months before for adaptation, notably in financial matters, to what they knew would be conditions of great confusion and restraint, to say the least. For example, in July 1939 Brigadier H St-J L Winterbotham,

Secretary of IUGG, proposed to its President (La Cour of Denmark)[7] that reserve funds be transferred to the USA. La Cour demurred on the grounds that it would be an unneutral act for a Dane to agree to. In the event, neutrality became almost a nonsense, as country after country was overrun, and its freedom of communication and financial flexibility nullified.

There were a good many scientists far from home, some unhappily in countries with which their own were at war. The free movement of scientists, an emotional ideal not yet distinguished from the refugee problem generated by the 1914 war and a constant concern of the League of Nations, had already become a question of harsh reality, to which the scientists of the countries opposed to the Axis powers had to give specific attention. Freedom to return to a homeland, freedom to leave a homeland which had become intolerable: both desires are exemplified in the personal lives of many scientists active in ICSU or its Unions at the outbreak of war. To go into the long tales of cruelty and deprivation alleged by all combatants about all opposed combatants serves no purpose here. The sum total adds up to the forced withdrawal of many good scientists from the service of their countrymen and the community at large.

Experiences of ICSU and the Unions

In 1939 forty-two countries adhered to ICSU, and still the same seven Unions: IAU, IUGG, IUPAC, URSI, IUPAP, IGU, IUBS. Adherence of different nations to these Unions followed no common pattern of adherence and management. For example, IUPAC could look to many chemical societies in different countries as representative national bodies of which there was no equivalent in other sciences. In each Union there was a different pattern of adherence to various Commissions, Associations or whatever title each Union gave to autonomous sections, often because special subject groups had existed longer than the Union which now brought them together. Notably, the USSR was not a member of ICSU but was a member of IAU, and had continued its long concern for geology. (It had hosted the International Geological Congress in 1897 in St Petersburg and in 1937 in Moscow.)

Finance was no tidier than it had ever been. Each Union had a schedule of dues, some countries paying faithfully, some lagging or defaulting. In spite of this there was in each Union and in ICSU itself enough liquidity and enough reserve for the essential service of communication to be carried on continuously.

The Second World War is for other books to describe. In the distant perspective of events now half a century old, it is easy to suppose a sudden change: it can be imagined as if a fluid continental Europe crystallised overnight into two rigid parts. It was not quite so. France was not at once subjected and divided.[8] A good deal of academic scientific effort was still possible for a while, even as far east as Strasburg, where the geodesists tried to carry on their work with an enemy very close. Italy was not a declared participant in the war until the invasion of Belgium, so its important place in volcanology could be (in principle) maintained. The subject did subside after a while, until revived after the war. The USSR, after the episode of the non-aggression agreement with Germany, was aloof from western affairs, although beginning the absorption of Baltic states that confused membership for a generation, as the active membership of the Baltic states and Finland was lost.

After the General Assembly in London in 1937 no Assembly was held until 1946, again in London. In the meantime the secretariat resided at the office of F J M Stratton, Secretary-General from 1937 to 1952, who kept the archives and funds of ICSU intact.

C Fabry was President from 1937 until he resigned in 1945; he was succeeded by the senior Vice-president H R Kruyt (Netherlands) until the next General Assembly in 1946, when J A Fleming (USA) became President. During the war years ICSU was kept viable if dormant. This held more or less also for most of its members.

There is one very important factor in the normal condition of scientific work, of which many non-scientists (and grant-making agencies) seem sometimes to be unaware, namely easy, continuous personal contact. This is true even within countries, let alone internationally. So in wartime, because of the transference of many scientists to war-oriented work, the progress of science in some subject areas was restricted and communication suppressed. In some areas it was stimulated, but it had to wait for the peace before it became generally beneficial.

IAU

The fortunes of the International Astronomical Union[9] during the war are typical of the mixture experienced by most Unions. Some work was carried on, some had to be abandoned. Some techniques developed for war purposes opened up new ways of studying unwarlike phenomena. (An example is the use of radar, the classic example of the adaptation of a new military device, being used in meteorology for detecting rain formation). Shared with other Unions was the experience that great gaps opened up with the loss of German expertise and the admired German commitment to comprehensive detail.

Time is of the essence: the aphorism can be drawn in to all manner of discussion. The Bureau de l'Heure continued its work on the comparison of time signals in Paris, work which was constantly being improved in effectiveness by new techniques.

IUPAC

The activity of the International Union of Pure and Applied Chemistry was reduced to a formal minimum. Chemists, perhaps more than other scientists, can move easily from the theoretical domain into the applied and industrial, and the 'art' of war had now become an industry. However, some pure research not cloaked in official secrecy continued to be pursued, and the Commission of Atomic Weights was able to publish a report in 1941. There was also prepared a table of isotopes in 1942. The Annual Table of Constants also appeared. These were all service functions, rather than contributions to fundamental chemical knowledge, but they show a tenacity of purpose.[10]

There were positive factors which played a part in continuity. Some of the work of revising data for publication was carried out in the United States under the direction of Professor H S Taylor.[11] Similar work was carried on by a refugee in England, the Belgian, J Timmermans.[12] Surprisingly, a working party on analytical reactions and reagents consisting of Van Nieuwenburg in Delft, Gillis in Ghent, and Wenger in Geneva managed not only to keep working together but even to prepare a substantial volume, published eventually in Basle in 1945.

URSI

To open a meeting in September 1946, URSI looked back on its own history, the early period being dealt with by Louis de Broglie, brother of the noted Charles-Louis de Broglie,[13] and the recent history by E V Appleton.[14] Appleton emphasised that, in wartime, in common with other scientific workers, the radio scientist had to work under restraint or in secret. But much that had been found had turned out to be of lasting value. This was especially true of radar, which had benefited by the design of new vacuum tubes and the development of new techniques of generating and examining radiation in the radio frequency spectrum. The accessible range of that spectrum had been extended into shorter wavelengths, which had helped with study of the atmosphere and had led to the development of new techniques of radio-meteorology. The study of the ionosphere had helped in the prediction of transmission conditions. The release from wartime constraints meant that stations could now be sited independently of strategic considerations.

Appleton looked forward to the restoration of the valuable URSIgram (described in chapter 4) so far as it continued to be useful. It had continued to operate well up to the invasion of the Low countries, but then could only keep watch. However, by following the dictates of the occupying forces about keeping people at work, it was possible to take care of the URSI office and papers in Brussels. All the same, in spite of economies, the URSI finances were in a pitiful condition at liberation, but funds had shown a gradual increase to useful level by the time Appleton reported.

IUPAP

IUPAP had been exceptional in trying out conferences with IIIC before the war, so, with the severing of communication with the League of Nations, it lost a useful resource. Like some others Unions it was forced into inactivity during hostilities. Its French General Secretary, Henri Abraham, was deported and died in captivity.[15] The X-ray crystallographers, who were prominent in the work of this Union, kept some research going and, after the war, helped to found the IUCr under the leadership of Lawrence Bragg.[16] The archives of the Union and a small bank account were kept intact.

IGU

The International Geographical Union geographers had run an excellent congress in Amsterdam in July 1938 and had put in hand a number of substantial projects (which were not entirely suppressed by the war).[17] Portugal was chosen as the country for a congress in 1942, but at the outbreak of war it was decided to reduce the international activity to the administrative work of the Bureau, and to encourage so far as was possible such work as could be carried out within each country.

The Secretariat remained at Louvain. In August 1940 the archives were removed to Germany, but were returned two years later, disordered and defective, but able to preserve some continuity.

The President (E de Martonne), in Paris, and the Secretary (M A Lefèvre), in Louvain, managed to keep in touch and to maintain a dignified independence. Although activity was inevitably at a low level, some dogged determination did produce results, notably the work on an international geographical bibliography, directed in Paris by Elicio Colin[18] and ready for the press at the end of the war, incomplete but very valuable. A cartographic bibliography was also prepared covering the cartographic documentation published in France and its colonies from 1940 to 1944 (including maps of countries other than France). The problem of financing publication remained, but the work had been done, under daunting conditions.

IUBS

'A vacuum' is how the 1946 report of the Secretary of IUBS described the war years. This was in part due to the same conditions as hampered other Unions, but IUBS felt rather more than other Unions the isolation of its General Secretary, Professor M J Sirks, of the Genetisch Instituut, Groningen. He was denied all foreign relations, and even 'undesired' contact with occupying forces. Before the war IUBS had made a great effort to draw the Germans into the Union, and therefore more firmly into the international community. They had just succeeded in getting three prominent, trusted German biologists[19] sufficiently interested in Germany joining the Union, when the outbreak of war nullified all the good that had been done.

The 1946 report of IUBS refers to one area of collaboration which it shared with IUPAP, but was utilised by few other Unions, namely work with IIIC. In the years 1937–9 they had the help of IIIC in organising three conferences on special themes (Phytohormones: Paris, 1937; Ionising radiations: Groningen, May 1939; Terminology in genetics: London, August 1939). It is difficult to make sense of the attitude of German authorities to these conferences. It is by no means certain where authority to control movement lay: the German Foreign Office, or the Academies themselves. Delegates attended the latter two conferences but were denied permission to attend the first. However, so far as IUBS was concerned, there did still seem to be some official German desire to keep in touch with international organisations.

IUGG

The International Union of Geodesy and Geophysics has left a very detailed account of its passage through the administrative perils of war.[20] Some of its experiences can no doubt be parallelled in other Unions, but one picture will serve. Its General Assembly met in Washington in September 1939, already uneasily. The declaration of war caused many delegates to pack their bags and seek passage home as fast as possible. In spite of this, a good deal of scientific work was done before the closure. It was decided, however, that no administrative decisions could be undertaken under such conditions. Within a short time normal communication between the countries which had been most prominent in IUGG's work became difficult or impossible. What could be done was to get some work published for the early years of the war at least.

IUGG was fortunate in the provision that had been made for the safeguarding of its funds in peace-time conditions. The Union and most of its Associations had their funds deposited in the same United Kingdom Bank. Subscriptions were suspended for the duration, but, so far as its means allowed, the Union made subventions in support of what work could be done by those Associations and other organisations to which it had access.

Death is no respecter of administrative tidiness, and the death of the IUGG President, La Cour, left only one competent signature, so a new trusteeship had to be created for the legal management of funds. Such considerations seem very far away from the high-minded pursuit of learning, but they have to be lived with if the work is to be done.

ICSU reviving

In war, the harshness of life and cruelty of death are intensified. Many people on both sides who could have added to our scientific riches died in battle, in attack on civilian targets, as prisoners, under torture or under deprivation of the common needs of life.

One of the problems that affected all Unions was the safeguarding of archives. An example is that of the International Association of Seismology. On the outbreak of war, the University of Strasbourg was evacuated to Clermont-Ferrand, taking with it the Association's library and records. Later, some of this material was deposited in a chateau in the Jura but was discovered by the Germans and taken to Jena. At the end of the war, negotiations had to be put in hand through diplomatic channels, not of the easiest, to try to recover them. Another loss was that of experience. Many of those who had time to give to a Union up to 1939 now found that with all the goodwill in the world they simply did not have time for it under the pressure of national demands for efforts towards recovery.

What of ICSU itself? Up to this point we have said little about the way ICSU was administered. It had always been run personally on lines inherited from the original association of academies system. Each Secretary-General used the secretarial facilities of his academy or of his employing academic institution. There might sometimes be hired secretarial help, minimal in amount, and professional accounting and auditing for the small amounts of money involved. This might seem to limit the scope of operations, but it had one advantage in the conditions of survival in and emergence from wartime conditions: the personal system was very flexible, so survival was sure and emergence swift. This could only be true, it hardly needs saying, if there were capable officers. The members of ICSU seem to have chosen well.

At the end of the 1937 General Assembly, the next was scheduled for Copenhagen in 1940. In the event it did not take place until 1946, in London. The constituent Unions still numbered only the seven already listed. National membership (then 39) fluctuated, as it always has, with the winds of political change.

The functions of Secretary-General of IRC and then of ICSU had been effected from the United Kingdom from 1919. The Royal Society had the status, the financial resources, and the accumulated expertise continuously to support an international office that IRC and ICSU needed. It did not limit its backing to its own Fellows. Since 1937 the position of Secretary-general of ICSU had been held by F J M Stratton, a man of immense experience in the administration of scientific organisations. He had been Secretary of the British Association for the Advancement of Science from 1930 to 1935, and Secretary of the International Astronomical Union from 1925 to 1935. ICSU affairs had to subside for a while but Stratton was ready to activate a meeting of the Executive Committee at the earliest possible moment after the war. The official address of the Secretariat was given as Gonville and Caius College, Cambridge. Stratton was elected to Fellowship of the Royal Society in 1947, but long before this the main support for the work of ICSU was activated by the Royal Society. By 1946 the Secretariat's official address was clearly stated as 'care of the Royal Society'. But there was more to it than a mere change of postal address. The whole scientific community had changed, in composition, in outlook, in habits, and in objectives.

Everyone was affected by experience of the habits of military collaboration on methods of communication. Secrecy and confidentiality had long been a habit in industrial science. The Patent system offered some protection for the rewards of inventiveness, but restraint in communication was even more effective in some kinds of work. War imposed the habit of restraint all round.

There are at least two sides to this, public and professional. One public result was the shock of exposure to new weapons, which was followed by a public interest in how the new results had been achieved. The magnetic mine, the flying bomb, and the rocket stand for stages in the intensity of that interest. The flying bomb was a deadly menace for the brief moment of exposure. Then the atomic bomb stunned one nation and cast a shadow over the future of every other. The experience of the prevalence of secrecy had a permanent effect on the public image of the scientist: whereas the scientist had been thought of as remote and beneficial, now he, and, increasingly, she, had sometimes to be thought of as remote but potentially harmful.

On the professional side the hampering of exchange of publication was a kind of intellectual malnutrition felt in every branch of science, and shared with the humanities. There were several aspects: the economics of publication reduced the amount that could be afforded by publishers, commercial or academic, who were, in addition, still suffering from the disturbance to their levels of staffing. Movement of publication from country to country was affected less by physical considerations than by slowness in the relaxation of exchange controls.

The end of the war brought no sudden alleviation. The opponents in the war, who had never lost their respect for each other's scientific prowess, could not restore the lines of printed communication which had been severed by hostilities. As if this were not enough, there was a new factor: the USSR and the Western Allies were soon at arm's length, attracted by common needs and repelled by political suspicion. The scientific communities on both sides quietly ignored political strain whenever they could. So ICSU was involved in a changed world. Support for the restoration of administrative effectiveness by a strong and respected national academy was badly needed, and the Royal Society had no superior in this respect.

A world at last at peace (or so it seemed) could not be a world at rest; there was new work to be done which included the creation of conditions in which science could flourish. ICSU had a larger part to play in world science than ever before. Much was happening and the way was forward, not back.

Chapter 6
A new beginning (1945)

From war to peace: the foundation of UNESCO

The change in ICSU's position and role in the scientific and cultural world was to be an integral and significant part of a world change.

The war which had been several wars came to several ends. But, long before those ends were in sight, political and cultural leaders had begun to formulate a successor to the League of Nations, not necessarily of the same form but with some of the same purposes. There were early discussions, such as the Atlantic Charter agreement between Churchill and Roosevelt (1 August 1941).[1]

The term 'United Nations' was used in a declaration by 26 nations in January 1942 supporting the ideas behind the Atlantic Charter. In August to September a conference met at Dumbarton Oaks in Washington to prepare proposals which became the source of the United Nations Charter.[2] In April to June 1945 there met in San Francisco a general conference which evolved a Charter for a United Nations Organisation. This came into formal effect on 24 October 1945 with the deposit of the first ratification (by the USSR) followed by those of a large proportion of the Charter signatories. Thus was the United Nations conceived, gestated, and born.

One of the most important elements in the reasoning for its creation was the belief that merely to unite nations was not enough. Other unities were needed. Clause 57 of the United Nations Charter therefore provided for the creation of several specialist agencies,[3] one to be devoted to education and culture. The idea did not originate in the San Francisco discussions. Some years previously (16 November 1942) there had opened in London a Conference of Allied Ministers of Education (CAME),[4] which continued its discussions until 5 December 1945; 18 governments took part. One of the results of this idea was the convening in November 1945, following a proposal by CAME, of a conference, to formulate the constitution of an educational and cultural organisation as envisaged in the UN Charter. Science did not feature in the original proposals, but the view that science should be included was forcefully expressed by some distinguished individuals, for example by Joseph Needham and Julian Huxley. This was accepted, and UNESCO (rather than UNECO) it became.[5] On 16 November 1945 the Constitution of UNESCO was signed by 37 countries, the Final Act by 41 countries, and, by 39, the Instrument setting up a Preparatory Commission, the body that was to do the effective work of getting the thing going. On 3 December 1945 the Executive Committee of the Commission held its first meeting, electing Sir Alfred Zimmern as Executive Secretary. Zimmern had devoted a large part of his life to the study and promotion of internationalism.[6] In the thirties he had been Secretary of IIIC and so was an obvious choice for the post. But times had changed, and the pressures of administration of the mid-forties were beyond even his great experience.

There were 15 seats on this Committee. One was left vacant for the USSR. Although it had been the first to ratify the Treaty setting up the United Nations, it did not join UNESCO until 21 April 1954.

By March 1946, Zimmern had to resign because of ill-health, and his place was taken by a universally respected biologist, Julian Huxley.[7] National Commissions began to be set up in member countries: between 13 June 1946 and April 1947 six national Commissions were set up, Brazil taking the lead.

In September the Commission moved its staff and office from London to Paris, occupying a hotel in the Avenue Kléber. The constitution of UNESCO came into force with the 20th ratification (by Greece, on 4 November 1946). The first General Conference of UNESCO took place from 19 November to 10 December 1946.

ICSU and the new political world

An event of great importance in the history of ICSU was the adoption, by UNESCO, on 6 December 1946 of provisional directives concerning relations with non-governmental organisations (NGOs). Only ten days later on 16 December 1946 the first agreement between UNESCO and a non-governmental organisation was signed, the NGO being ICSU. However, this good event did not take place without much discussion. Having reviewed the UNESCO background, let us go back over what was happening inside ICSU while the UNESCO discussions were consolidating UNESCO itself.

By the time the Fourth General Assembly of ICSU met in London in July 1946, the discussions on the formation of UNESCO had progressed far enough for ICSU's relations with the proposed new body to be the subject of fertile debate about the approval of agreements. Another personality, one who has now been prominently visible in the international field for over half a century, had joined in the discussion: Joseph Needham, Head of the Natural Sciences Division of the Preparatory Commission, who, with a distinguished career in biology already to his credit, had spent a large part of the war in China as a scientific liaison officer, and held views on international cooperation which were based on a more up-to-date kind of experience than was the fortune of most people to enjoy. He was able to help ICSU see how its role could extend into the future with a far wider coverage of nationalities than ever before. He prepared a memorandum for general use on his ideas, which is an excellent survey of the problems of post-war co-operation. His words, based on some central ideas in this memorandum, addressed to an ICSU Committee on Science and its Social Relations had a considerable effect throughout the General Assembly.[8]

The intimate relation between ICSU and UNESCO was well founded in these formative years by individuals who remained part of ICSU's life for years. Outstanding among these is Pierre Auger,[9] who had worked in Canada with French colleagues on atomic physics. Helpful in his later work was friendship with Henri Laugier, Assistant Secretary-General of the United Nations, who expressed his own attitude to international science by advocating the creation of international laboratories. The first of such laboratories to appear was CERN in 1956. Auger succeeded Needham as Associate Director for Pure and Applied Sciences. He acted as a voice for ICSU inside UNESCO and as a voice for UNESCO inside ICSU

The address of welcome to this 1946 ICSU General Assembly by Sir Robert Robinson,[10] President of the Royal Society, was no mere formality, but expressed a conviction of the vital need for balance. He spoke briefly but to the point on 'the noble ideals quoted in the Draft Report of the [Science and its Social Relations] Committee'. While respecting their admirable intentions he warned of the difficulties of putting them into effect. 'Carried to their logical outcome, the precepts of some enthusiasts suggest that loyalty to science must transcend loyalty to king and country. In others a political paramountcy is claimed which goes far beyond anything the community will tolerate. I mention these matters only to beseech you to apply to concrete proposals the acid test of practicability.' To many thoughtful people nowadays it must seem as if we are still trying to find that balance between the high-mindedness of the Committee's proposals and Robinson's advice to reflect on practicability.

To place ICSU in its contemporary background at mid-1946 we can use some passages from the presidential address by Professor H R Kruyt.[11] He not only provided a portrait of ICSU as it stood, with its strengths accepted and its weaknesses exposed in detail; he described the attitudes he thought it should adopt in facing changed circumstances.

He reminded the meeting that this International Council had to consider a world which in many respects was fundamentally different from that of 1937. The weaknesses were twofold: in the first place the constituent bodies of ICSU were so strong in themselves that there was no obvious role for ICSU in their specialist fields. In the second place, since ICSU was a confederation of national academies and Unions, it could be no stronger than its constituents. ICSU could therefore be most effective in helping to form new Unions, and in initiating work in fields which fell outside or between the fields of existing Unions. (Kruyt was far-sighted: this is a good summary of what did in fact happen.)

The Executive Committee had been obliged to report the loss by ICSU and by Unions of many strong individual members, some by natural causes, some of them by the accidents or malice of war. Among others were Charles Fabry (of the Fabry–Perot interferometer),[12] Marconi (most celebrated example of the scientist-entrepreneur), Arthur Eddington (astronomer and interpreter to the public of relativity theory),[13] Henri Abraham (geographer who died in a gas chamber),[14] Henry Lyons (geophysicist, reorganiser of the South Kensington Science Museum, Secretary of the Royal Society). Good men all: the miracle is that it has always seemed possible to find good successors.

Another 'casualty', a paper casualty, was the agreement signed in 1937 between ICSU and the IIIC International Institute for Intellectual Co-operation, a body which looked anomalous with the demise of the League of Nations, and whose existence was obviously vulnerable.

Although the first post-war General Assembly could not meet before July 1946, the time since the end of the war had been used in the gathering of old threads and with the spinning of new ones. The first new thread began with the receipt of a letter from Sir Alfred Zimmern (who was still in the Preparatory Commission at the time the letter was drafted), reporting a Resolution of the Preparatory Commission to this effect:

> That the Preparatory Commission of the United Nations Educational, Scientific and Cultural Organisation be requested by this Conference to instruct its Executive Committee to consult with the International Council of Scientific Unions on methods of collaboration to strengthen the programmes of both bodies in the areas of their common concern, and

> That the plans thus formulated be reported to the first Conference of UNESCO, with recommendations for a suitable working arrangement with the International Council of Scientific Unions.

A committee studied the draft with Joseph Needham, on 29 May 1946. It was discussed by the ICSU General Assembly on 22 July, and an agreed modified form was adopted by the first General Conference of UNESCO in November 1946.

ICSU had not long to wait to see what sort of cultural world it had to live in. From this time on the work of ICSU and that of UNESCO were so intertwined that one is inclined to forget that UNESCO was not the only world organisation to possess and wield a scientific arm. WHO, WMO, IAEA, and FAO, for example, had considerable scientific interests embodied in their own constitutions.

ICSU could thus look at its world in two manners: that of broad goodwill, as indicated by UNESCO, a manner which those concerned hoped would lead to the general adoption of science as an integral part of the lives of all peoples. The other manner was professional in substance, that which matched the specialties of the Unions, areas of science limited in scope but rich in detailed content. Differences between these two styles there might be, but the character of the difference was to change. UNESCO took, and takes, the whole of human life as its province. ICSU was initially constrained by the limitations of the older Unions, in number and in scope. Breaking out of the bounds of those limitations was to be one of ICSU's (and the Unions') most important contributions to world science.

Against the background of the UNESCO *démarche*, the General Assembly wrestled with questions of revival and advance. Among the main issues was that of old Unions and new Unions. Were the existing Unions up to new tasks and were new Unions needed or desirable? There were bodies not connected with ICSU which had adopted the title 'Union'. This suggested they might well be candidates for membership of ICSU. On the other hand, although good communication was desirable, it was not a good idea to let every small special-interest group acquire the administrative status of a Union. (The cynic may reflect on the position in the United Nations where many groups small in territory or population or gross national product have acquired nation/voting status.) As so often happens, a flexible, pragmatic attitude looked like being the best answer, and so it turned out.

The next step was the drafting of an agreement with UNESCO. The 1946 debate brings out the complexity of the situation in which ICSU found itself, much wider in scope than had been the case only six or seven years previously. ICSU was faced by many more people, many more tasks, potentially a much higher level of financing, and the existence of other international organisations with a concern for science. After much debate a wording for an agreement[15] was settled, the most important clauses of which were the first two:

1 UNESCO recognises the international scientific Unions as providing a natural and appropriate form for the international organisation of science and the International Council of Scientific Unions as their co-ordinating and representative body.

2 The International Council of Scientific Unions recognises UNESCO as the principal agent of UNO in the field of international scientific relations.

The word 'relations' seems vague but it was, in fact, a source of strength to ICSU, since ICSU was to be concerned primarily with scientific action, and it would be able to define its work in far more specific scientific terms than UNESCO.

The 1946 General Assembly still occasionally looked over its shoulder at the past. In attendance at the meeting were representatives of IIIC which still existed although its days were numbered. On 23 July there came to an end an agreement which had once promised to cement relations between ICSU and the realm of the humanities. It had failed because of a clash of styles. As we saw in our account of the origins of IAA, a single multi-disciplinary body had been tried in the early years of the century. This had disappeared with the 1914–18 war. The League of Nations had set up its own International Committee on Intellectual Co-operation to act as the consultative organ. However, the Council of the League needed, like any such body, not just a consultative, but an executive, agency, which the League was not organised to create. The French government had acted on behalf of the League and set up under French law a body which did have executive powers: the International Institute for Intellectual Co-operation. In addition to working in the fields of education, literature, and philosophy, IIIC had tried to operate in the some areas of science, interesting itself for the most part in philosophical generalities, but also taking part in some joint efforts on specific scientific subjects (e.g. with IUBS). In 1937 it had signed an Accord with ICSU intended to consolidate their spheres of interest and action, but by 1946 it was clear that, while the role of ICSU had been confirmed by the setting up of UNESCO, the very comprehensiveness of the title of UNESCO was enough in itself to show that the separate existence of IIIC could no longer be justified.

In a letter of 16 June 1946 the Director of IIIC[16] 'denounced' (the English word reads harshly: the tone of the letter was, in fact, gently courteous) the Accord and asked ICSU to agree. A Establier,[17] for IIIC, explained that the cessation of the Accord was

intended to facilitate new arrangements between UNESCO and ICSU. The President
of ICSU responded with the valedictory warmth the letter deserved.[18]

Science looking forward: reports of unions

In July 1946 UNESCO was not yet formally established (this was still a few months
away), but all the reports of Unions presented at the General Assembly of ICSU were
expressed in terms of a new life for science and a new public view of science.

Sir Edward Appleton, elaborating on the formal report of URSI, spoke of Radio
Science as being a subject which lent itself readily to international collaboration. 'The
laboratory of the radio-physicist is the world itself, largely because of the existence of
the ionosphere.' He referred to the radio-physicist concerning himself with other fields
of natural knowledge: meteorology, atomic physics, and astrophysics. Recent developments fostered by war (importantly radar) might be of permanent significance. There
could hardly have been a better field to illustrate a general effect, but it was not unique.

E de Martonne, in concluding his survey of the work of the IGU, pointed out the
difficulty of restoring the old contacts and making new ones. The next Geographical
Congress could hardly be on the same scale as of old, but eventually the level of
research would be restored. The chemists reported preparations for immediate meetings of its Bureau and forward planning of a Congress. The physicists and the geographers were doing much the same. It had taken less than a year for all the Unions to turn
their backs on the past and face the future.

But there were troubles generated by the persistence of the economic strains of war.
Jacques Hadamard[19] moved for action on exchange control and its effect in hindering
movement of publications. Although other speakers mentioned some local arrangements, more of a charitable nature than systematic, there was evidently a general
problem. It was decided that this was not ICSU's level of responsibility. It was a matter
for governments and should be put to UNESCO to consider along with other problems raised by exchange control.

Science and social relations

The last two days of this General Assembly (23–4 July 1946) were largely taken up with
discussion of the resolution proposed by the Committee on Science and its Social
Relations

The main proposals of the Committee were: a plea for the elimination of the domination of science by military secrecy; the seizing of the opportunity for international
collaboration in applied science presented by the prospect of atomic power; the maintenance by the scientific community and its members of an attitude of 'frankness, openness and integrity' in the application of their work to the common good; and the
promotion of the public understanding of science. In the event, the General Assembly

adopted a resolution along these lines. (Of the hopes expressed in the resolution, the last, on public understanding, seems to have taken practical form in the later years of the century particularly with the development of new media involvement.) To get the contemporary background to the Committee's reports one needs to look at its appendix, a survey of recent publications urging the commitment of the scientist to peaceful pursuits. It all seemed very bright and hopeful at the time, but a science-exploiting world had yet to show its power to defeat good intentions.

Atomic warfare, not unexpectedly, had dominated the preliminary discussions. No other activity had been so demanding of secrecy, a fact which was deplored. The level-headed Needham pointed out that secrecy can sometimes be necessary. The references to secrecy in the final resolution were strong; worded so as to assert that excessive, dominating, secrecy would be not only destructive of scientific productivity but counter-productive of the military intention. (This discussion was, of course, fission-bomb oriented. The catastrophic character of the fusion bomb moved the moral debate out of the local consequences of military action to that of world-wide survival.)

A V Hill[20] went further into science-based perils. He urged the General Assembly to take a sombre look forward: while recognising that biological weapons had not been used in the recent war, he still asked for a ban on biological warfare. The discussion recognised the analogy with nuclear energy. In both cases, great benefits might arise from discoveries that, wrongly motivated, could be put to the wrong use. The Committee eventually came out with a good, strong, acceptable motion,[21] (i) urging international agreement on applications of nuclear energy, (ii) stressing the need to consider the threat of other means of warfare than nuclear, such as biological, (iii) acknowledging the duty of the scientist to serve the community as professional and as citizen. There were thanks all round and everyone went home to face an uncertain scientific future.

Other new international organisations

So far, the only new United Nations organisation we have considered has been UNESCO, but there were others which bore a scientific aspect, such as the Food and Agriculture Organization (FAO), the World Health Organisation (WHO), the World Meteorological Organisation (WMO).

To choose one, for illustration: the World Health Organization was established in 1948 with the general aim of promoting international co-operation in measures to improve health by all necessary means. Its forebears were a Health Organisation of the League of Nations (set up in 1923) and an even more remote predecessor, an International Office of Public Health (with headquarters in Paris) which had as its objectives the control of epidemics, quarantine measures, and the standardisation of drugs. WHO, however, had a much wider mandate. It was expected to promote the highest possible standards of health for all people. The definition of health was firm and strong: 'a state of complete physical, mental and social well-being and not merely the absence of disease or infirmity'. In keeping with the general objective of the United Nations it was held that good health was conducive to peace and security.

The responsibility of WHO which interests us most here is that of disseminating information on scientific advances in medicine, including the encouragement of consistent manufacture of useful drugs, old as well as new. Its triumph in recent times has been the eradication of smallpox, but so brilliant an achievement could hardly have been foreseen at the time of its establishment. One could give many examples of cross-fertilisation in its interests, but one will do: a developing interest in air pollution as a health hazard, which helped create the basis for the realisation that the atmosphere is a precious and limited resource for mankind. Another, for ICSU more important, example, is the World Meteorological Organisation (WMO) of the UN with which, over the years, ICSU developed strong co-operative links through the Global Atmospheric Research Programme (GARP, 1967–80), the World Climate Research Programme (1980), and, most recently, the Global Observing Systems.

A mere string of acronyms can show how various were ICSU's eventual relations: ICSU eventually enjoyed consultative status with UNESCO, WMO, IAEA, ITU, FAO, WHO, UNEP, ECOSOC, UNDRO, Council of Europe, IDNDR, CSD, WIPO.

New Unions

ICSU had developed with only seven Unions. By the time of the 1946 General Assembly it was possible for everyone to look at the way science had changed since 1937, the last time it had been possible to meet for a discussion of fundamentals. The discussion had two interesting aspects. In simplified terms: What is a Union for? What are the conditions in which it is desirable that a Union should exist?

Gradually there emerged another question, again not expressed in explicit terms but implicit in the discussions which led to the creation of many new bodies over the subsequent 40 years: what other kind of organisation or mode of action is needed when the straightforward Union system appears inappropriate or inadequate? This issue was to turn out to be as important to ICSU as the others, perhaps even more important in the long term.

New Unions

The question of new Unions came up early and forcibly in Kruyt's Presidential address on 22 July 1946. He quoted one of the ICSU Statutes: 'to co-ordinate the national adhering organisations and also the various international Unions', and said this seemed rather theoretical, since organisations cannot be co-ordinated, only their activities. The success of the Council therefore depended on the liveliness of the adhering bodies. An important aim of the Council should be to establish Unions in fields where this liveliness fails, and ICSU still lacked several Unions. For example, although there were Unions of Geodesy and Geophysics and of Geography, there was no Union of Geology. There was an international congress of geology that met every three or four years, but there was no organisation to promote activity between congresses except what was required to see that the next congress took place. That congress organisation was usually set up in the country where the next congress was to be held, so there was no stimulus to international action. If a nominated President or Secretary should happen to die (not to speak of war) there was no machinery of continuity. If, on the other hand, the congress committee were to be a local commission of a Union, the Bureau of ICSU could always guarantee continuity and moreover provide a link between the successive congress committees.

Kruyt then said that, although this might look like a mere matter of administration, it was more than that, because another clause of the Statutes referred to ICSU's duty to 'direct international scientific activity in subjects which do not fall within the purview of any existing international association'. ICSU could not direct activity in geology because there was no representation of geologists in the Executive committee or in ICSU itself so long as there was no geological Union adhering. Moreover, it was not possible to call into being joint commissions requiring co-operation between other scientists and geologists if there was no geological representation. Kruyt compared this imagined situation with the real situation in respect of the *Ionosphere,* where there was a joint commission between IAU, IUPAP, IUGG, and URSI. There were joint commissions on *Rheology* between IUPAC, IUPAP, and IUBS, and on *Oceanography* between IUGG and IUBS. But for Oceanography one missed a Union for Geology, and for Rheology one missed a Union for Mechanics.

It was to be a long time before a Union of Geology was in fact founded (1961), but other Unions were formed to extend the scope of ICSU membership. By the time of the next General Assembly (1949) three more had been founded, on quite different bases. The International Union of Theoretical and Applied Mechanics, (IUTAM) representing an old established and long-practised science, was already in process of formation in 1946 and was formally established soon after. The International Union of Crystallography represented a new science, a leading figure in its formation being one of the early founders of the science, Sir Lawrence Bragg. Informal talks during an international meeting at the Royal Institution in May 1946 had planned a Union, and prepared for the publication of an international journal. The Union was admitted to ICSU in April 1947.

The International Union of the History and Philosophy of Science (IUHPS) represents yet another mode of foundation. There had been a long succession of international conferences of history of science and of philosophy of science since the beginning of the century. The historians founded their Union in 1947 and adhered to ICSU. The International Union of the Philosophy of Science IUPS was founded in 1949, but had not been admitted to ICSU by the time of the 1949 Fifth General Assembly. (The two joined forces in 1956 to act as two divisions of one Union.) IUHS had already developed several scientific sections and was to multiply them as time went on. Since many Unions have members with an interest in the history of their own science, Joint Commissions with some other Unions were also created. Enough to say that the way IUHS later merged into IUHPS is characteristic of the evolutionary character of many Unions.

These three examples are enough to show that there has been no one way of forming a Union, nor one way of developing its work.

It is clear from the remarks of Kruyt (on, for example, the need for a Union of Geology) that those who were hoping to put ICSU at the forefront of international scientific activity were very conscious that it was not yet ready to make any claim to ubiquity. However, limited though it might be, it was the only body of its kind, as was apparent to the Preparatory Commission of UNESCO. That ICSU was given a unique formal relation with UNESCO is a tribute to the far-sightedness of the founders of UNESCO in acceding to the 'S' in its title. They were persuaded to see the need for such an extension, and were therefore the more easily convinced soon afterwards of the likelihood that ICSU would meet much of UNESCO's, and the world's, needs in this area.

It was going to take time, not only for ICSU, but for all the other new and reconstituted organisations of the post-war world, to clarify their purposes, to establish priorities, to find the right people, and above all to set up that machinery of staffing, finance, and administration without which no organisation has any real future.

Administrative structure of ICSU 1949

By the time of the Fifth General Assembly (Copenhagen 14–16 September 1949) the administrative set-up of the Council had developed from its early IRC beginnings to the following pattern.

The ruling body was the General Assembly which was required to meet every three years. On it both national members and International Unions were represented, by delegates appointed by adhering organisations. Voting powers varied: decisions on scientific questions were taken by a majority of votes cast by all delegates present. On administrative and non-scientific matters the vote was taken by adhering organisations, each national member having one vote and each International Union three votes or two votes according to whether they were ranked as General or Specialised.

The distinction was one of the supposed breadth of the subject covered by the Union: for example chemistry was considered General, geography Special. It is clear from the heat of the debate on these distinctions that they were doomed to be superseded at some time in the future. A two thirds majority was required to amend the Statutes. The official languages remained English and French.

The General Assembly was responsible for electing from members of the adhering organisations a President, two Vice-Presidents, a General Secretary, and two members of an Executive Board; each General Union appointed two members to the Executive Board, and each specialised Union one member. The President and the Vice-President held the same ranks in the Executive Board, which also included the retiring President.

The central administration was in the hands of a Bureau, which consisted of the elected members of the Executive Board. The legal seat of the Council was at Brussels, Belgium. The administrative headquarters remained (for the time being) wherever the General Secretary resided. Titles of offices were sometimes informal in the early stages. For example, the word *Bureau* was used as shorthand for the group of principal officers later known as the Executive Board.

This looks very tidy on paper, but like so many schemes which have grown out of informal agreements it had eventually to be changed in the light of the growth of ICSU in a post-war world.

Chapter 7

ICSU in a post-war world

The changing picture of world science

It would be a pity to allow our study of science in a post-war world to be clouded by the confusion of nations in conflict, of enemies in imagined victory or defeat, or allies in friendship or rivalry. These factors are all there, but we need to maintain such detachment as we can.

To examine post-war ICSU we should look at the science which was its *raison-d'être*. What does war do to, or for, science? These questions embarrass some people, but they will not go away. Any attempt to find simple answers to them arouses so much debate that we might be wise to confine ourselves, so far as we can, to reporting. Naturally there were a great many attempts to estimate the position of science from the point of view of all kinds of group: learned societies, professional organisations, universities, political pressure groups. If we start with an American example of such an estimate, it is because it could well be read now with a world's eye.

'The endless frontier'

We can profit from the significance of the title of this famous 1945 report[1] prepared by Vannevar Bush for the people of the United States to whose President it was addressed. The frontier had always played a central role in the American people's thoughts about their status as a nation. From the time the inhabitants of the early colonies began to push out beyond the confines of their first settled land, the frontier was that line beyond which lay hope and hazard. To confront hazard and to fructify hope needed vision and hard-headed organisation. The benefits were unpredictable but could well be enormous. Vannevar Bush could have offered no more imaginative image of the prospects for science for the whole world than that of the frontier as seen by the American people. While those who looked to the future of science in other countries may not in fact have used just this metaphor, it well describes the thinking of many, scientists and non-scientists. Many of them worked on matters which do not call for the elaboration of theory, but all of them have been trained to be aware that their fields of knowledge extend to the far frontiers of the intellect. Very optimistic views of a supposedly beneficial science were held by some 19th- and early

20th-century writers. These were devalued by the manifestly technical basis of the 1914–18 war.

It is not easy to distinguish between changes which would have come about anyway and those which were markedly accelerated or initiated by war. By 1939 mass production was already the leading philosophy of manufacture. New materials were already coming into use, and even newer were being sought. The techniques were available to combat wartime shortages, for example, of rubber, by the discovery of new techniques of synthesis.[2] Wartime change in transport was for both the good and for the bad. Free movement in transport was restrained by war while, at the same time, the balance between methods of transport was changing very rapidly. The movement of objects and people by air and by road was made faster and more reliable, while railways, one of the greatest achievements of 19th-century technology, declined in potential in some parts of the world because, with capacity overstretched and equipment undermaintained, they lacked the kind of research and development that went into new rivals. They declined almost to extinction in some parts of the United States. They remained one of the great unifiers of India.

Medicine benefited from advances in chemistry and biology, so that between 1939 and 1945 one could be said to have entered an era of a successful opposition to infectious disease. Another change in medicine was the general adoption of the idea that medical problems could be attacked as a matter of routine by seeking for aid from the physical and biological sciences and even from engineering. One outstanding example is the complex history of the discovery and manufacture of penicillin, which is made up, however, as we have remarked, of uncertainties and rivalries as well as of co-operation and generous sharing of experience and equipment.[3]

Communication of information changed radically. The common reliance on radio for rapid dissemination of information altered not only the daily habits of ordinary people, but also the machinery of government. It was not all acceleration, however. Some of the basic principles of public service television had been established and were ready for widespread application as war began, but had to wait for a while before television could embark on its eventual social ascendancy.[4]

With the widespread destruction of beautiful cities and productive industries from Coventry to Dresden and Hiroshima, what might have been thought of as normal life remained remote. In every nation, even those which did not have to endure battle on or over their territory, the disturbance to the economy of daily life through the scarcity of essentials and of replacements made hardship widespread. It was because of this shared endurance that, in spite of shortage and disturbance, the scientific communities in most countries were able rapidly to renew their co-operation on a peace-time basis, albeit a new one.

The heritage of war included some distinct benefits, such as the earliest computers to be used seriously. There were also many curious but welcome benefits. For example the accidental observation in wounded aircrew that fragments of Perspex (Lucite) lodged in an eye were well tolerated led to the development of optical implants which have been a blessing to cataract sufferers.[5] Rocket propulsion, however, seems the most massive advance and is two-faced. It eventually opened up nearby and distant space to exploration and service; from space also was cast the shadow of violence.

None of us can resist using hindsight at times. The new insecticide DDT was a blessing at the end of the war in helping to control insect-vector borne disease and in controlling crop pests. The damage done to other forms of life was not adequately foreseen, but was to be a component of eventual ecological passions and policies. It is the sort of thing which intensified the old debate about the moral responsibility of the scientist.

The computer provides a notorious example of the perils of prediction: when the first successful computers were in action it was predicted that in a decade or two there might be a few large-scale computers in industrial use. In fact, in that time computers became widespread. The revolutionary personal computer (coming into widespread use in the 1980s) was to confound yet later predictions. Soon after the war great hopes were held out that nuclear energy would provide for all energy needs, cleanly and without limit. It was not long before the inadequacy of that view was exposed. It was not so much that provision of nuclear energy was not as easy as had been hoped, but that its production generated new international concern for promoting safeguards against those of its consequences which were unwelcome, such as dangerous radiations and hazardous waste.

The century had opened with a revision of methods of considering heredity and evolution, using the examination of whole organisms and their characteristics. By 1950 the scope of the study deepened, through the application of biochemistry to the components of the cell. It was not long before X-ray crystallography took it much further. The elucidation of the double helix structure of DNA has become one of the classics of scientific discovery, not only dramatic in its events but hugely influential in its opening up of new fields of science, pure and applied.[6]

This last example highlights a paradox: the smaller the natural entity being studied the bigger becomes the apparatus needed, and the more complex the organisation needed to bring different strands of enquiry together. The lone scientist of an old romantic tradition almost disappeared. The few survivors were mathematicians or observers of creatures in the wild. Even as early as his first graduate research studies the scientist learned to expect his future career to lie in working in teams. That main career was now increasingly likely to be spent in a large institution, engaged in work converging in diverse ways on some central problem or depending on the availability of some large shared experimental resource.

In addition to the academies, new supplementary research organisations were formed like CNRS (Caisse National de Recherche Scientifique from 1935, and then Centre National de Recherche Scientifique from 1939) in France, and the TNO (Toegepast Natuur-Wetenschappelijk Ondersoek) (Applied Scientific Research) and ZWO (Zuiver Wetenschappelijk Ondersoek) (Pure Scientific Research) in The Netherlands.

There were also new regional bodies, especially in Europe. The European Coal and Steel Community (1952) recognised the primacy of economic and industrial factors in international affairs, but this body was succeeded by the European Economic

Community of 6 nations (which added to its number until it reached a powerful 12 and considered other possible members). For this a Science Directorate was established in 1973. In parallel with this, regional research institutions had emerged like the European Organisation for Nuclear Research (CERN) (1952), set up by UNESCO, the European Molecular Biology Organisation (EMBO) (1963), and the International Institute for Applied Systems Analysis (IIASA) (1987) which helped maintain scientific liaison between East and West.

The range of institutions increased. Fundamental science was spreading from the academic world into the industrial: the example already set by men like Carothers[7] was to be a common way of life. And the proliferating scientific institutions were filled by more and more scientists, writing more and more, to the point where it was impossible for anyone to read everything written in his own speciality, let alone in any wider field. No one could relax comfortably into quiet contemplation of an area of which he could be confident he knew the limits. The frontier was a common frontier as it had never been before, making conditions in which ICSU reacted to novel opportunities.

Evolution of ICSU's structure

For 25 years the IRC/ICSU office had lived a nomadic existence, making camp at the desks of the successive Secretaries. It had a short period of near independence in rooms in Tavistock Square, London. Then it was able to set up house.

Until 1952 money matters had been dealt with by the Secretary. From then on there was a separate officer, a Treasurer, who could be chosen for his skill in financial matters, to share the burden and focus on the problems of an increasingly large budget.

The sacrifice of free time by officers of ICSU, and by these successive Secretaries in particular, was very great, and now with a programme of expansion in view it was clear that this state of affairs could not continue. Just as the conduct of science was changing, so the way it was reviewed and co-ordinated had to change. The creation of UNESCO was a recognition of this fact. It was a piece of good fortune for science at large and for UNESCO in particular that ICSU already existed, with a structure which made it ready to play an immediate part in the metamorphosis of the scientific world.

An improvement in ICSU's position was effected in 1947 by UNESCO providing office accommodation and pay for a Liaison Officer. The post was first occupied by A Establier[8] who had worked with the IIIC and had been in close touch with ICSU. He had space in the first UNESCO building (the Hotel Majestic in the Avenue Kléber). It was the beginning of a new phase in ICSU's life, a phase in which a permanent professional Secretariat was to become increasingly important and to provide a model for the management of science policy-making, particularly after 1953. Following Establier's resignation in 1949, Ronald Fraser[9] became Liaison Officer. After six years in Paris, on 1 November 1953, the ICSU–UNESCO Liaison Office was moved to the Royal Society in London to work in closer touch with the Secretary General (then A V Hill), the title of the post being changed from Liaison Officer to ICSU Administrative Secretary.

Space was limited at the Burlington House rooms occupied by the Royal Society which was planning a move, and in 1958, at the invitation of the Royal Dutch Academy of Sciences, the ICSU Secretariat moved to larger accommodation in the Palais Noordeinde in The Hague. We can leave these administrative changes for the moment because the time the Secretariat spent in The Hague marks the transition from what we can think of as the immediate post-war period to a longer period of ICSU development.

Finance

The changes in personnel were dictated in part by financial changes, some internal, some a reflection of world events. It is always difficult to get an idea of the relative value of money at times past, even in the recent past. Figures are on record for the dues charged by ICSU and by the Unions and the subventions paid by UNESCO to ICSU for general purposes and for specific projects, but the thousands of dollars of these early years seem very small compared with the millions involved in analogous functions in the early 1990s.

For example, the Report of the ICSU Finance Committee in 1949 showed the total travelling expenses paid for the years 1947 and 1948 to be £1412 and £1194 respectively. The average total cost of subsistence allowances for *all* the members of a committee per committee meeting was £70. Limitations on the number of meetings of committees which could be supported were proposed.

The main accounts were kept in sterling in England, but there were also accounts in US dollars, French francs, Belgian francs, and Dutch guilders. Although the UNESCO grant was apportioned to each of the bank accounts appropriate to the individual purpose for which it had been agreed, the accounts were kept in a form which showed the state of ICSU's own self-generated income and expenditure. The balance in 1949 was favourable but only just. 'During the past two years we have lived just within our income with a small surplus of £65 to spare.'

At this time, as during the entire existence of ICSU, the national members provided most of ICSU's subscription income: in 1948 national subscriptions totalled £1059, Union subscriptions only £135. At the same time, subventions from UNESCO for specific purposes began considerably to exceed the funds ICSU raised on its own behalf.

One topic in the 1949 General Assembly debate shows the gradual way UNESCO policy as well as ICSU policy evolved. The matter at issue was: what should UNESCO support: organisations or projects? To put it another way, should it be expected that UNESCO should support an organisation as a whole or only individual projects being carried out by those existing bodies? However, could the *creation* of a new body devoted to some new field have the status of a *project*? In the end it was agreed to put to UNESCO as criterion that it should support new projects, which would include the formation of new necessary bodies, and also initiatives (and the capacity to initiate), by existing bodies. In time so many new projects were created by ICSU and other bodies associated with UNESCO that the question became irrelevant.

In 1952 the Seventh General Conference of UNESCO authorised the Director-General to establish an International Advisory Committee on Scientific Research. The intention was to promote international co-operation between national councils and centres of research where there were fields of common interest. The UNESCO Executive Board expressed this in more practical terms at its 36th session: 'to advise the Director-General on research in the natural sciences and related matters in the programme of UNESCO and on the promotion of international co-operation in scientific research'. The Committee was set up in December 1953. The President of ICSU and representatives of two other bodies having the same world scope as ICSU (e.g. CIOMS, the Council for International Organisations of Medical Sciences, and UATI, the Union of International Technical Associations) were ex-officio members. Direct representation of certain Unions on this Committee came later, in 1962. It was being recognised that the changing character of science demanded flexibility in the organisation of its support.

National representation

At the first post-war General Assembly in 1946 ICSU had 39 national members.[10] The General Assembly reports of 1949 and 1952 quote 43 and then 44. There was a core of old members who had been involved since the days of IAA and IRC, but there was some fluctuation among the increasing number of new members. On the whole there was steady growth from now on. The kind of national representation varied, as had always been the case, since the definition of types of member in the Statutes remained accommodatingly broad. It might be an academy, it might be a government department having scientific responsibilities and acting in a non-political manner. Whatever the character of the applicant, membership was not automatic: an application had to be examined by a committee and any recommendation approved by the General Assembly.

Membership required that dues be paid and some nations did not pay up very quickly. When it was proposed in the Executive that such nations should be expelled, tolerance prevailed. The scientific communities in some countries had little chance of getting the ear of finance ministers. They needed ICSU more than ICSU needed their small subscriptions. There were new nations to be considered, some coming into existence peacefully, some passing through phases of violence. When the calm of some degree of political stability had prevailed, one by one new nations sought to work within ICSU. No two were alike.

The international political scene on which ICSU found itself an actor from 1945 onwards, was shaped by the UN, by UNESCO, and by changes in the balance of power among the larger nations. Much of this was quite new but ICSU still had also to face some old problems in new guises, the problems IRC had met after the 1914 war: persuading old enemies to make friends, recognising the existence of private quarrels between old and new nations, patiently persuading suspicious nations to recognise ICSU's value to them.

As of old, Germany was a major problem, but now a twofold one. Although much of its scientific potential had been diverted into war purposes (as had happened in all the belligerent countries), its scientific pre-eminence could not be denied, and it was important that it be brought into the international field as soon as possible. The mathematicians were particularly aware of this, so great had been the German contribution to the subject. But it was not the unified pre-1939 Germany, nor the many-principality pre-Bismarck Germany. Its traditional weakness of internal division (which had already been evident in the difficulties over the formation of IAA) had now reached a new form with the division of Germany, after an interregnum, into BRD and DDR, in 1949. They were two nations on paper, but they had a common intellectual heritage. ICSU had to accommodate them.

Japan was also undergoing a political transformation, but this was part of an internal change which was eventually to bring it into the forefront of applied science. For the time being, bad western conscience about the atomic bomb, and uncertainty about Japan's potential made relations with its scientific community difficult.

The USSR presented a new kind of problem. Its internal political system had caused it to be viewed with dislike and apprehension by some other powerful nations, a feeling which was grimly reciprocated at a political level. Although there was a good deal of friendly and productive communication between Russian and other scientists it was not until 1954 that the USSR joined ICSU. During the so-called Cold War the interaction with the Russian member was fully maintained and proved to be of special significance (which is referred to in a later chapter in connection with IGY and COSPAR).

Of particular interest was a widespread community which was the source of a new nation. Israel came under special scrutiny in the scientific world, because of the worldwide recognition of the special position of the Jewish intellectuals who had left Germany. Some, but by no means all, returned to Israel when it was established as a nation-state in 1949 and helped to create a new scientific community. It became a member of ICSU in 1952.

India can claim an ancient scientific tradition, and had enjoyed a growing academic life during the British Imperial period, particularly after the foundation of three universities soon after the Indian Mutiny. By the end of the century work was being done of Nobel Prize stature (the work of Raman in spectroscopy).[11] A national academy capable of representing its considerable scientific community already existed when the Imperial rule ended in 1947, but this had to establish itself as the representative of a new post-partition India which was different in many ways from the old. The other new post-partition nation, Pakistan (which suffered its own partition with the separation of Bangladesh), now also had to identify the credentials of its own scientific community.

China is different again. It too has a glorious scientific past, widely known now in the West after the work of the devoted Joseph Needham,[12] but its contributions to modern science are of more recent times than those of other great nations. It has another old tradition of warring factions, but for much of the first half of the present century it was subject to wars within wars. At the end of the 1939–45 war confusion settled into a condition of what the outside world saw as two Chinas, but which each of those two claimed to be only one. The island of Formosa, with a Chinese population, had been administered by Japan until 1945. Two million mainland Chinese, under Chiang Kai-Shek, joined the 20 million inhabitants, and a state of Taiwan claimed independence. The United States patronised it, keeping chilly relations with mainland China. Republican China (whose membership was clearly very desirable) would not associate with any body which recognised Taiwan. (This situation was eventually resolved first by the recognition by UNESCO of the People's Republic of China as the only representative of China and by the United States recognition of the People's Republic of China in 1979.)

The early position can be summed up this way: in the reports of the General Assemblies of 1946 and 1949 the representative body for China is given as the *Academia Sinica, Nanking* (the capital of the Nationalist government led by Chiang Kai Shek. In the report for 1952 it is given as *Academia Sinica, Peking* (the capital of the Chinese Communist government). China and representation of Chinese science was a problem for the UN, for UNESCO, and for ICSU for a long time afterwards. (A solution adopted by ICSU and some Unions eventually was the appreciation that academies might sit down side by side even when governments could not.)

There were more new national members than new Unions. Before 1939 ICSU could expect to correspond with a modest number of academies with similar historical origins. Then only a few of the academies of wealthier nations could afford to support the cost of participation of their nationals in the honorary posts which maintained the work of ICSU and the Unions. With the expansion of ICSU's membership and the increase in its UNESCO-sustained funds, the expenses of scientists from poorer countries could be met, with the great benefit of calling on talent from all sections of the world community.

The attendance at General Assemblies was more and more varied, but it followed that like was not always speaking to like. On academic issues it might not always matter, but in financial questions some representatives might have to refer back to one kind of body, say a ministry, while others referred back to an academy. The old IAA had found it had problems of lack of uniformity. In spite of a great difference in structure ICSU was no better off in this respect. With the growth of science at large, the academies or science ministries now represented larger numbers of scientists, but the concept of equality of nations embedded in the United Nations constitution percolated down and replaced the old IRC idea of subscription and voting by population. The voting systems used by ICSU and by Unions had to be adjusted to a scientific world the components of which did not match the components of the political world. In both, diplomacy was the key.

The voting-subscription system adopted by ICSU was mixed. Each national member had one vote, but could elect to subscribe at one or other point of a scale of subscriptions. The classification of the Unions as Special or General (with different voting rights) was to be varied later, with all Unions eventually having the same status.

Unions

Soon after the revival of peace-time activity new Unions were added to the seven founders (IAU, IUGG, IUPAC, URSI, IUPAP, IGU, IUBS). The stimulus might be an independent body opting for changed title as Unions. Or it might be a new creation, stimulated into existence by the possibilities held out by Union status (e.g. IUTAM). Another might be the products of a new discipline (IUCr). At the General Assembly of 1949 there were three more Unions (IUCr, IUTAM, and IUHS). At the GA of 1952 the total number had grown to 11 with the revival of IMU.

In most countries members of academies were an élite, each member expert in a particular discipline, attending the ICSU meetings with a brief to think and act in terms of the whole scientific effort of his country, on the one hand, and on the other acting as a member of ICSU for the benefit of ICSU as a whole. On many issues delegates behaved like members of the world scientific community, discussing matters on their own merits, thus eliminating differences in membership category. Often an academy representative would also be a member of a Union delegation, which could make considerable demands on diplomatic skills, especially those of the President.

This description of the composition of a General Assembly might convey a picture of drab dullness and formality. Anyone who has ever attended one will know this is far from the truth. Few gatherings can be as lively and diverting, so full of variety and challenge. The one ingredient missing was ill-will.

Associates

There were thus scientific Union members and national members, but there was as yet no obvious place for membership by another kind of group, neither national nor confined to one discipline. Some of the nations, old as well as new, in Latin America, the Pacific, and in Africa may have appeared on the surface to have less to offer in the way of scientific potential or authority in proportion to their populations than the scientific leaders, but some were already involved in the new studies that were becoming prominent: in particular they were actors on an ecological stage which was to be increasingly important to world science and to ICSU. Some had already developed regional affiliations. For example, a Pacific Science Association[13] had existed since 1920 and could bring not only interest, but also experience. A new kind of association with ICSU seemed to be needed, and this eventually came into operation in 1968 with the creation of a membership category of *Scientific Associates*.

There were other factors which made ICSU face up to a different world from that in which it had been born. In addition to the bodies initiated by the United Nations and UNESCO there were new regional international organisations being added to those already in existence, some of them devoted to, or at least having an interest in, science and its patronage. These were growing in Europe, in the Pacific and in Latin America.

With each of these entities – new Unions, new nations, new regional bodies – ICSU established a *modus vivendi*, or a suitable form of membership. Similarly it did not limit itself to one rigid way of dealing with scientific problems, but set up new kinds of body, just as the experimental scientist devises, as needed, new apparatus. This is to be seen in the variety of new projects which soon became a major part of its activities.

New programmes

Let us look at examples of a new situation, of a new scope, of a new flexibility, and of a process of evolution. (In doing so from the point of view of ICSU, it will become apparent that in different ways and to different degrees UNESCO often played some part in them. Some UNESCO initiatives were developed into working programmes by ICSU; sometimes ICSU proposals were adopted by UNESCO.[14])

A new situation was created by the launching of the first artificial space satellite by the USSR in 1957. This was not of itself a cause of the opening of a new era of co-operation; the preparation for such an era had already been got under way. However, the fact that the USSR had signified by its membership of ICSU that it was disposed to provide a balance to a world scientific effort meant that planners were a great deal nearer to being able to think in global terms, and the Russian satellite showed that the global planners might well have a new physical instrument at their command.

A new scope was defined in the planning of a very important new project, which, at first began merely with the idea of following a good historical precedent. It changed course and style in the light of new conditions and new desires. The precedent was that of the International Polar Years (1882–3 and 1932–3). In 1952 informal discussions about the possibility of organising a further IPY led to its formal adoption by the Joint Commission on the Ionosphere, with later support by IUGG, IAU, and URSI. Later in the year the Sixth General Assembly of ICSU agreed to organise a new Polar Year to build on the knowledge and experience gained in the earlier ones. However, a study of the new objectives showed that confining the scope to the Polar regions or even to high latitudes would have neglected opportunities to pursue questions in oceanography, the problems of Antarctica, and the potential of new techniques in aid of research.

The Polar problems had gradually turned into global problems, best described as Geophysical in the most comprehensive sense. International Geophysical Year (IGY) it became, a project so important that a separate chapter is needed to describe it. The wide scope of the discussion revealed the need to have another programme devoted to the oceans, so SCOR (Scientific Committee on Oceanic Research) came into existence during the same planning period (July 1957).[15]

SCOR is an example of the transformations which can take place in both role and method of an international initiative. Its initial purpose was to 'further international scientific activity in all branches of oceanic research'. Its first major project was novel in that it aimed to explore an area which had been little studied so far, namely the Indian Ocean. ICSU took a leading part in the organisation and co-ordination of the work of this International Indian Ocean Expedition (IIOE). Responsibility eventually passed into the hands of UNESCO's Intergovernmental Oceanic Commission in 1962,[16] SCOR thereafter filling the role of general scientific advisor to the Commission. SCOR derives its strength as advisor from its programme of meetings on techniques and special area problems, and working groups which examine developments in methodology.

The growing flexibility of ICSU's management illustrates the unpredictability of scientific change in the inauguration, as another outcome of the IGY discussions, of COSPAR (Committee on Space Research). The satellite was the new instrument: it was clear that there would be a great deal of investigation using it. This would clearly generate as great a variety of questions and methods as any other new field of inquiry. So the need for co-ordination was self-evident. The IGY Rockets and Satellites Conference in Washington DC in October 1957 urged the importance of continued research following the International Geophysical Year. After a recommendation of the IGY Committee in August 1958 in Moscow, ICSU set up COSPAR[17] at its General Assembly in October 1958.

In estimating the evolution of the ICSU vision we should remind ourselves that what is mere fancy in one generation may become reality in the next. Well into this century the idea that we might contaminate unknown worlds with our own bacteria would have seemed mere science-fiction fantasy. It was now a manifest possibility and a committee (CETEX, Contamination by Extra-terrestrial Exploration) was formed to devise safeguards, such as seeing that what was sent into the unknown was sterile.[18]

All the same, the reality of space research is different from heroic science fiction, which seldom takes account of the massive scale and tedious minutiae of the ground support and administration. The value of COSPAR lies in its capacity to make its complexities intelligible to any participant, and to provide that service continuously. We might say this of any of the international bodies we are concerned with.

This recital of new Committees (and there were many others coming to birth) is characteristic of a change in the style and manner of scientific work in the post-war period. It is true that there had been a great many special groups before the war, even back into the 19th century, but they had originated, on the whole, as separate and independent entities.[19] The new style international scientific committees were growing out of a common belief in the effectiveness of inter-relation between the sciences, in other words of interdisciplinarity. The new committees would become one of the most important rationales of ICSU's existence as they were designed to address international interdisciplinary problems which no Union and no academy could have tackled alone.

No one believed that any one committee would last for ever. One might be initiated with an eye to a limited lifetime, made up of planning, action, review, and final publication. Another might develop into a service system so that, although it need not go on for ever, it had a good deal of continuity. They were all related in a manner which confirmed the wisdom of the analogy of ICSU as a family.

Chapter 8

The free conduct of science

The committee as watch-dog

No one owns a magic carpet. To many people who do not have to pursue the life of science it must seem ideal: the pursuit of new knowledge wherever one needs to go. In fact the freedom to practise science is as hard won and as hard to preserve as any other kind of freedom. ICSU has been deeply concerned with the problems of freedom ever since it was founded, and since 1963 has had committees devoted to the study of principles and to action in individual cases. The principles of freedom had been set out clearly enough in the United Nations *Declaration of Human Rights* in 1948,[1] but their observance was not mandatory: those who respected it were themselves free to take such action as they wished and could contrive; those who chose to ignore it did so without sanction. It might seem that restraint on freedom of movement of a scientist was a very small matter compared with the suffering of false imprisonment and torture. This is certainly true, but it so happens that the freedom of movement is an essential ingredient in the progress of science, in general as well as for the individual scientist.

In his address as retiring President to the Second General Assembly of ICSU in 1934[2] (read for him in his absence by Henry Lyons) George Ellery Hale said:

> The purpose of the International Council and its associated Unions is to promote co-operation in research irrespective of nationality. We welcome to our meetings the man of science in all countries and we appreciate the opportunity to join with them in the pursuit of our common object.

The years that followed saw political changes all over the world which produced barriers and hindrances to the practical operation of these ideals. The ideal of free movement was sometimes confused by a surge of the involuntary movement of refugees, financial insecurity among displaced scientists, and by the suspension of normal social life through war. Added to this was confusion as to who was a traveller, who a migrant, and who a refugee. There were many local aids to scientists on the move like the Society for Visiting Scientists in London in which A V Hill was an important figure.[3] These could only do a little and were short-lived. All the same the larger ideal was not lost sight of, and, in 1958, the General Assembly of ICSU went far beyond what it had done in 1934, when it just tacitly accepted a Presidential personal view. It now supported a policy resolution on political non-discrimination, more complex than Hale's personal view, and therefore worth quoting in full.[4] Thus:

> The General Assembly,
> In keeping with the purely scientific character of ICSU approves the following statement:
> 1 To ensure the uniform observance of its basic policy of political non-discrimination, the ICSU affirms the right of the scientists of any country or territory to adhere to or to associate with international scientific activity without regard to race, religion or political philosophy.
> 2 Such adherence or association has no implications with respect to recognition of the government of the country or territory concerned.
> 3 Subject only to the payment of subscriptions and submission of required reports, the ICSU is prepared to recognize the Academy, Research Council, National Committee, or other bona fide scientific group representing scientific activity of any country or territory acting under a government, *de facto* or *de jure*, that controls it.
> 4 Meetings or assemblies of ICSU or of its dependent organisms such as its Special Committees and its Joint Commissions should be held in countries which permit participation of the representatives of every national member of ICSU or of the dependent organisms of ICSU concerned, and allow free and prompt dissemination of information related to such meetings.
> 5 ICSU and its dependent organisms will take all necessary steps to effect these principles.

A new discussion was triggered by the attitude of the North Atlantic Treaty Organization (NATO) which had been set up in 1949. It imposed a virtual ban on the provision of visas for entry into Nato countries for scientists who were East German nationals.[5] At its meeting in Prague in October 1962 the ICSU Executive Board decided to ask its officers to take action in the general matter of visas. By 1963 a Working Party was able to present a report and recommendations which were the foundation of an ICSU mechanism which has survived, with modifications of constitution and mandate, to the present day. Successive resolutions became more bold and positive. The language of the 1963 resolution forming the Standing Committee on the Free Circulation of Scientists (SCFCS) is different from that of the 1958 resolution. It reads:

> The Assembly *considering* that ICSU has a declared policy supporting free international collaboration among scientists, but
> *noting* with regret that there are still parts of the world where difficulties exist in the free passage of scientists,
> *reaffirms* the declaration of 'political non-discrimination' adopted by the VIII General Assembly of ICSU in 1958,
> *resolves* that, in holding ICSU meetings, and meetings of ICSU Scientific and Special Committees and Inter-Union Commissions, the Council shall take all measures within its powers to ensure the fundamental right of participation, without any political discrimination, of the representatives of every member of ICSU concerned and of invited observers.
> *recommends* that this policy be adopted also by the Unions adhering to ICSU for all their activities,

> *invites* the ICSU national members also to follow this policy, and
>
> *requests* the Executive Committee to create a standing working group to assist the Officers' Committee to find solutions to various specific problems associated with the implementation of this resolution.

The first Chairman of SCFCS was the Norwegian, Nicolai Herlofson, Professor of Electron Physics in Stockholm. He expressed the policy of the Committee in these words: 'The Committee Members, from the UK, France, Soviet Union, Czechoslovakia and the US with a Scandinavian Chairman, understood that their main task was to remove as many obstacles as possible for free international scientific collaboration. The instruments were continuous safeguarding, analysis and interventions with a low publicity profile . . .'

Herlofson was Chairman from 1963 to 1972. His successors have been T Caspersson (1972–1982), Inga Fischer-Hjalmars 1982–1991, and S Helmfrid from 1991 onwards. The Committee functioned without an Executive Secretary until 1972, when Olof Tandberg took on this role which he has filled with enthusiasm until the time of writing. In its early stages it dealt successfully with problems arising from the partition of Germany which still presented many international problems of great difficulty. It managed to get entrance visas from the Allied Travel Office in Berlin for many East German scientists to enter several West European countries, and eventually managed to eliminate the need for documents from the Allied Travel Office, normal visa arrangements taking their place. What was, and remains, necessary is the need to be clear that the scientist wishing to travel does so as himself, not as a representative of his government.

The change from hopeful desire to practical action, which one can see in the difference between the 1958 and 1963 Resolutions, is evidence of the way ICSU was becoming conscious of its power to intervene in the conduct of science at large, instead of just within a closed field of specialisms. Science was no longer to be considered as being just for scientists. The International Geophysical Year, with its shared concern for the human condition, is an example of a change in outlook. SCFCS survived and survives by a paradox: by insisting that scientists should be treated as individual free people it makes their science the responsibility of everyone.

ICSU was, to some extent, ahead of UNESCO in this: Unesco's own elaborate and comprehensive *Recommendation on the Status of Scientific Researchers* came in 1974,[6] but before then ICSU had already found that good intentions need to be backed by hard work. Herlofson's report to the ICSU General Assembly in Madrid in 1970 explained the Committee's position and the degree of its effectiveness:

> The Committee works by correspondence. When complaints about visa refusals or excessive delays of visas are received, it solicits information, usually through its members, clarifies the position and draws attention to the ICSU policy as expressed by the decision of the 10 General Assembly in 1963. Any official action is reserved for the ICSU Officers, who are advised by the Committee for the Free Circulation of Scientists.
>
> The number of cases considered is less than 10 per year at present; the tendency is towards a decrease. Officials within some of ICSU's national adhering organisations

state that the existence of the Committee and the continued reference to the declared ICSU policy are of value when negotiating with their national ministries or visa authorities. It is repeatedly emphasized that the aim of ICSU is to work in a constructive way towards reasonable compliance with its free circulation principles; the aim is not to embarrass the governments of countries where visa difficulties may have occurred.

The low profile remained, but the number of cases did in fact increase as ICSU widened its membership and its functions. The ICSU Executive at its meeting in Erevan in October 1963 unanimously agreed 'to request the Unions and their subdivisions to keep a record of visa refusals and abnormal delays, the Unions to exchange such information annually, and the Unions and their subdivisions to take this information into account when selecting the location for future meetings'. This was more easily said than done, but 'a realistic standard procedure' was slowly developed. By 1970 the movement of citizens of the German Democratic Republic had been very much eased, but problems remained which often emerged in the preparation of meetings by Unions.

Some problems which look like Free Circulation problems are intractable by reason of their political character. The same 1963 Executive Committee got involved in a discussion of the case of Professor Samueloff. He had been returning to his native Israel from an IBP meeting in London when his plane was hijacked. He was held by the Syrian authorities. The debate at Erevan was heated. One of the opposing points of view was that the retention of Samueloff had arisen in a scientific connection and that it was therefore ICSU's business to take action. The other view was that, since Syria was not a member, ICSU could take no action through normal scientific channels, and, ICSU, being a non-governmental organisation, could not act itself through governmental channels. It could only refer the matter to the United Nations through such diplomatic channels as were open to it. In the end the debate was closed by the President (V A Ambartsumian) on the grounds that it was a political issue and therefore inadmissible.

Tandberg has described SCFCS as a watch-dog,[7] but it took some time to grow to maturity and learn to bark at the right time. However, the Committee had the corporate wisdom to realise this and to declare that it was not enough on its own. Herlofson put it this way in a report to the ICSU General Assembly in Bombay in 1966: 'ICSU is one component in a group of forces that continually acts towards greater freedom of travel. A uniform procedure by the Scientific Unions and the other ICSU organs when arranging meetings could add to the effective pressure exerted by the scientific community.' The advice implied in this observation was soon to be incorporated in a firm policy statement and recommendations aimed at the universality of science.

Experience was showing, however, that new working arrangements were needed on a continuous basis, not just as individual cases arose. So when Caspersson became Chairman he developed a system of archives which has been a valuable and necessary tool for the Committee's work. In particular it provided a basis for interaction with other aspects of ICSU's work, and the spinning of a network of communication with

Unions. Freedom of movement questions often first come to light within Union activities, and it is often a Union which will find its work impeded by restraint of movement or an individual being prevented (as in examples quoted later) from taking part in some meeting or study activity. The official SCFCS reports to the ICSU General Assemblies sound as if it were all very easy and that the SCFCS just operated a routine procedure. The truth is very different. Anyone who has been allowed to look at the archive and study any individual cases realises how much work, by correspondence, by phone, and by personal contact, often by a number of people, can go into relieving the distress of any one individual. It is not only the individual who can benefit. From a Union point of view SCFCS often works to keep up the efficiency of a Union's own work, relying as it so often does in having the right people in the right place at the right time.

A good deal more than this crystallised in due course. Out of the study of movement problems came a strengthening in ICSU's attitude to the organisation of international meetings on the principle that ICSU is, and must be seen to be, consistently non-political. This is not a negative attitude. On the contrary, without being openly defiant, ICSU has been prepared to express its concern for universality continuously and doggedly, even going beyond UNESCO in its insistence on freedom from arbitrary governmental restraints. There has had to be the utmost care in dealing with restraints which affect scientists within their own countries. This was spelt out at the 1973 General Committee in Leningrad where the report on Free Circulation included comment on cases where an application by a scientist for an exit visa to attend an international meeting had been followed not only by refusal but by dismissal from post. ICSU could not intervene in what was an internal political matter for the country concerned, but it could do one thing. Since the political decisions were unlikely to have been taken by the national ICSU representatives of the country concerned, ICSU could draw their attention to the case, with evidence and documentation. It would then be appropriate for those national members to view the situation within their own political system and their responsibilities to it. It would be up to them to use their own channels of communication to impress on their political authorities the unfortunate impression liable to be created by their actions, and the consequent damage to the efficiency of their own national scientific effort.

Every national scientific body develops its own way of coping with its government, and it is to the national body that ICSU must address itself, not to the government, if a question of government policy arises. This was the position in 1966 when the Executive Committee had to face up to the exodus of highly qualified scientists from universities in Argentina. Even if they found employment elsewhere their students were left at home without teachers. Harrison Brown suggested that if scientist had to leave they could be encouraged to go to other Latin American countries rather than to the United States. However, it was agreed that ICSU could intervene so long as it kept to communication with the national representative scientific body and did not attempt to address the government. The outcome is clouded by the subsequent history of the political and economic instability of Argentina over the next two decades.

Where it is not in danger of adopting a political attitude, ICSU can demonstrate its attitude to a visa policy in several ways. An example arose in connection with the meeting of the Executive Board in October 1967, which had been scheduled for Athens but was eventually held in Rome. When asked for an explanation the Secretary-General (K Chandrasekharan) said that when ICSU had received an invitation to hold a General Assembly in India, the authorities had been asked to provide a written assurance that there would not be any difficulties regarding visas for the participants, and that this written assurance had been received. When ICSU was invited to Greece a similar assurance had not been received from the Greek authorities. ICSU therefore chose to go elsewhere.

There was a long debate on general questions of ICSU's structure which had begun at the Thirteenth General Assembly in Madrid in 1970. The reports of SCFCS had always, up until now, been received with approbation, but at the Executive Board in Ottawa in September 1971 one item is recorded with a different tone: J Coulomb said that at the present time the Committee was only duplicating and circulating complaints. Whether this was criticism of a failure of the Committee to be more resolute or a masked plea for the Committee to be given more power and authority is not clear. Whatever may have been Coulomb's intention, it is recorded that the Committee was enlarged.

That its reputation was good on the whole, in spite of Coulomb's veiled criticism, came out at the end of a long discussion, centred on the migration of scientists, at the ICSU General Committee in Leningrad in September 1973. Even the formalities of official reporting cannot conceal the heat that this debate engendered over criticism of the political philosophy of the Soviet Union, and over questioning of the reality of ICSU's non-political stance. In the end a sufficient distinction was drawn between matters of migration and of free circulation for the SCFCS to receive general approval for its work in what had to be a limited field. Migration was held to be a matter for another committee, a Standing Committee for the Safeguard of the Pursuit of Science, which was set up at the General Committee in Ankara in 1974.

In 1976 SCFCS was able to put down firmer administrative roots. The Royal Swedish Academy of Sciences granted the Committee the use of office space, telephone, telex, and meeting room, as well as the services of personnel for the management of the Archives. Olof Tandberg, the Executive Secretary, set in hand the further development of the Archives which had been initiated by Caspersson by adding to existing records material he collected from Unions, Committees, and Associates relating to the activities of SCFCS since its creation in 1963. The Archives needed a fairly elaborate indexing system: by countries, Unions, meetings, citizenship, entry visa refusals, exit visa refusals, entry visa delays, exit visa delays, refusal to participate in meetings in one's own country, emigration wishes, imprisonment. It was intended that any cases submitted which fell outside the mandate of the Committee should be forwarded to relevant bodies such as the new Standing Committee on the Safeguard of Pursuit of Science. Meanwhile the number of cases annually dealt with had risen from 10 in 1970 to about 40 from 1976 onwards.

A guide to free circulation

Hitherto the ICSU principles on Free Circulation had been made known along with ICSU's general principles and policies, without ever being concentrated into some clear presentation on its own. In 1976 SCFCS produced a new document which has continued (revised from time to time) to be the main instrument with which ICSU declares its policy and practices in this field of the Freedom of Science. This 'blue book', *Advice to Organizers of International Scientific Meetings*,[8] sets out procedures which it recommends should be followed by any ICSU-linked body when planning such a meeting, a major concern being the granting of visas, both for entry and for exit.

It has changed over the years. There is big difference in style and manner to be seen in the successor document, the current *Universality of Science*. This expresses itself in firmer terms, encourages organisers to be resolute in insisting on freedom of movement in attendance at proposed meetings, and sets out all the ICSU General Assembly Resolutions which have contributed to ICSU's present stand.

A fundamental question concerns the principal organiser (Union or whatever) which is advised to make clear that 'no obstacles will be put to the granting of visas to *bona fide* scientists who wish to participate'. It also states that no one should be prevented from participating in international scientific meetings because of 'race, religion, political philosophy, ethnic origin, citizenship, language or sex.'

The phrase *bona fide* is often repeated in the ICSU documents. It is a necessary protection, in principle, from the abuse of hospitality by persons entering a country for supposed scientific purposes, but in fact with the intention of engaging in non-scientific activities. It also provides a safeguard against the intrusion of political agents into the scientific community. This last safeguard is very difficult to operate. The question of *bona fides* is thus double-edged. ICSU claims freedom for the scientist, but he or she must be the real thing. As illustration: in 1989 two persons from Iran applied for visas to attend the IUPAC Congress in Stockholm. Requests sent to them to provide evidence of their scholarly and professional status produced no response. Did they just change their minds? Or were they unable to show that they were genuine scientists and so gave up the attempt to travel? What matters is that, if they were in fact trying to carry out a deception, the SCFCS procedure circumvented it. SCFCS was not trying to police a procedure. Had they responded positively, SCFCS would have done all it could to expedite their visas.

ICSU has been wise (one could say cynical) enough to advise any international organisation, on whose behalf a local organiser is acting, to reassure itself independently (from ICSU or SCFCS) of the past record of the country concerned with regard to the provision and delivery of visas. Protestations by officials of the host country are not enough. The organisation hoping to be welcomed should find out from other organisations what their experience of visa treatment and efficiency has been.

Getting a visa is never as simple as it looks. Time flies for the individual, who may get the impression that time seems unimportant to an official in a visa office. Where a visa is delayed or withheld for no apparent good reason, ICSU has a machinery for action and appeal to the country concerned. If appeal fails and the case appears to justify going on, disapproval can be expressed by means ranging from publication of a protest to cancellation of the meeting, removal of the meeting to another place, or recommendation to all member bodies that invitations to hold meetings in the offending country should not be accepted until it mends its ways. Invalidation of voting procedures can embarrass the country concerned.[9] Every country likes to look well. Prestige matters.

The officers of SCFCS were always realistic about politics. While they had to express principles in straightforward general terms they were well aware that the patterns of government in Africa, Asia, and Latin America differed, often radically, from those based on the European and North American traditions, that in each continent there were wide variations. They also realised that in freedom to move there was a danger that free movement might deprive a country with no long tradition of scientific work of men and women it needed for its own development. The 'brain drain' was the other side of the coin of free movement. It had to be taken into account as a counterbalance to a blind commitment to movement of any kind. Awareness of the problem served to reinforce the sense of common service that lay behind the international enthusiasm of the leading figures in ICSU. The individual and the community had to be thought of together. This is so important an issue that it became a factor in the operations of COSTED (Committee for Science and Technology in Developing Countries) which was set up in 1966, but SCFCS has, on the whole, tried to avoid becoming involved in cases other than those affecting individual movement from one country to another for some identifiable scientific purpose.

The continuous quiet inquiry of SCFCS exposes another problem arising out of the fact that the scientist in our modern society is generally a hired man. If he is in a government service he owes a duty of commitment to keep secret what his government decides is in the national interest. If he is in the pay of an industrial or commercial concern he earns his keep on the condition that he will not deprive that concern of its commercial advantage by prematurely revealing any aspect of his work that might have a profitable outcome. If he is an academic he may claim that his thoughts are his own and belong only to the community of his calling, that he may tell whom he likes of what he has discovered or devised. However, it may be asserted in some countries that its economic system is such that the state is the direct employer of the academic. Then the situation is not so clear, and the extent to which the scientist has relinquished a degree of freedom in order to feed his family or care for her children has to be carefully resolved.

Freedom seems therefore to be always qualified, to some degree or other. It is not enough for SCFCS to think of a scientist only as a scientist. He or she is always other things as well, entitled to a freedom which is always relative. As SCFCS has developed its awareness of these complexities its resolution of individual problems has become quicker and more effective.

Sometimes it is not possible to categorise the problems simply, particularly where a government or form of government is in a state of flux. China, for example, has been a long drawn out source of puzzlement and dispute, coupled with admiration for achievement. The China which is more correctly referred to as the People's Republic of China (PRC) and Taiwan would both be desirable members of any international organisation, but to bring representatives of both of them to one table seemed at one time impossible. Several bodies, ICSU and some of the Unions, have found solutions by invoking the rule that their members are not governments but groups of persons with the status of academies. When Chiang Kai-Shek took his forces to Taiwan he took the Academia Sinica with him to Taipei. The People's Republic of China (PRC) did not form a replacement academy but set up a different organisation, the China Association for Science and Technology (CAST). Eventually it proved possible for the bodies to attend ICSU or Union meetings together as separate academies of what both thought of (in different ways) as one country. In 1982, after prolonged negotiations about the definition of 'national member' in the ICSU Statutes, both academic bodies became full ICSU members. This success culminated in 1988 when ICSU held its Twenty-Second General Assembly in Beijing, where the delegation of Taipei participated and photographs of its members appeared on the front pages of daily newspapers. Relations with the PRC have not settled down to the calm condition that might have been hoped. The events of Tiananmen Square were regretted outside the PRC. Within the PRC they have been explained in terms of official policy, and those responsible for work of internal transformation regret what appears to be a lack of understanding and sympathy on the part of the 'foreigner'. The SCFCS policy towards countries which present political problems to those outside has been one of patience and guarded openness, trying always to keep up good relations with their scientists as scientists. It has slowly borne fruit.

The Committee at work

SCFCS developed its work slowly under the guidance of a succession of Chairmen together with the Secretary who expanded its scope over the years. It is difficult to provide a measure of overall success. Of course the existence of SCFCS and its well-known activities has also had considerable preventive impact from which non-ICSU organisations have also profited. In the case of entry visa applications it has been possible since the archives were set up to count cases. The records use the phrase 'positive outcome' to signify that the desired but obstructed movement did eventually take place. Since 1985, a positive outcome has been achieved in more than 80% of cases. In addition to action in individual cases, the information accumulated in the enquiries and negotiations gradually helps policy discussions, so that ICSU's position is strengthened in negotiations in which freedom is only one factor. It must also be said that failure to obtain a visa often turned out to be due to the applicant not having correctly carried out the procedures laid down by the country concerned whether for entry or exit. The applicant must play his part in getting the machine to work on his behalf.

The discussion may reach the highest officials. In 1982 the Australian government made difficulties over the provision of visas for two very senior USSR scientists for attendance at a Conference of IUB in Perth. SCFCS raised objections to this exclusion, and a series of exchanges began which culminated in approaches to the Australian Prime Minister himself. Some other USSR scientists had been admitted without difficulty; the objection seemed to be that the two in question were so senior that their non-scientific positions had to be taken into account. They looked more like government officials than straightforward scientists. In the end the confrontation was inconclusive, the Australian government claiming that there had only been a misunderstanding about the time needed to process a visa. The excluded USSR scientists had been prevented from exercising their right to vote in an important IUB election. IUB produced an ingenious face-saving solution. The voting at the General Assembly was closed but not counted. The urn was sent, under seal, to the USSR delegates at home. They recorded their votes, the urn was returned and the votes were counted. Throughout they had behaved with dignity, asking that the problem be resolved as quietly as possible. A similar situation does not appear to have arisen since.

Ironically it was the Australians who were among the most active in protesting at the attitude of the Korean government towards a number of intended participants in the International Union of Nutritional Sciences (IUNS) Congress in Seoul in 1989.

At the ICSU General Assembly in Ottawa in 1984 the Committee had to report several contrasting cases together. One concerned 54 cases of visa refusal for attendance at the International Geological Congress in Moscow, and another concerned 3 refusals for a Biotechnology Symposium in New Delhi. They were reported in exceptional circumstances, since another case had occurred: visas had been refused for two members of the delegation of the USSR to the General Assembly. This was held to be without precedent: no previous case of refusal of a visa to an official delegate to an ICSU General Assembly had been recorded. This was an embarrassment to the Canadian hosts since they had given assurances when presenting their invitation to ICSU that no problems would arise. The fault did not lie with the Canadian National Research Council (which was praised for its efforts to resolve the situation) but with the Canadian government department responsible. There could hardly be a better illustration of the recurrent problem facing SCFCS: put crudely, any country wants its nationals to be free to be admitted to a meeting abroad. The same country may want to be free to deny admission to anyone it considers undesirable. SCFCS will evidently never lack for work.

In 1988 there occurred an important challenge to ICSU's principles of the universality of science, the occasion being a Scientific Conference of the International World Ocean Circulation Experiment (WOCE), held at UNESCO in Paris. WOCE was a joint activity of two intergovernmental bodies (WMO and IOC), together with ICSU's SCOR. The conference was announced to the members of each of these. Interested scientists from each of two non-UN countries (Taiwan and South Africa) signed up and attended. Each participant was at the conference in an individual capacity. After the opening of the meeting the observers from the UNESCO Member States from Africa complained that it was illegal to hold such a meeting at UNESCO with scientists from

South Africa, and a walk-out of all Africans, subsequently supported by the partici-pants from other developing countries, was threatened. The ICSU Secretariat then began a series of intense negotiations to resolve the problem. ICSU's position was clear: either the conference continued with the scientists from South Africa, who were there in their personal capacities, or it should be moved to another venue in Paris which had been offered to ICSU. The negotiations with UNESCO's legal staff and with the delegates from Africa were, in the end, successful, and the meeting continued with all the participants at UNESCO. The argument that all were there as scientists, coupled with a public statement from ICSU's National Member in South Africa that it fully endorsed ICSU's stand on non-discrimination also in their own country, was instru-mental in obtaining the positive result.[10]

The background to the work of SCFCS is bound to change. The political scene in South Africa has undergone a transformation which many believed could never come about. The changes will have an effect on the conduct of scientific work in South Africa and consequently over the whole of the continent. That South Africa could not be con-sidered in isolation was one of the points brought out in an important debate in the Twenty-third General Assembly in Beijing in 1988. This debate began as a simple act of reporting, and turned into a variety of expressions of feeling of a depth which even the cold language of a conference record cannot conceal.

A formal presentation of a report triggered off an intense discussion of the meaning of universality. In 1987 there had been set up a President's Working Group on the *Interpretation and Practical Applications of ICSU's Principle of Non-discrimination in the Context of the Contemporary World*. Its deliberations had been as short as its title was long-winded, and it might be supposed it could do no more than repeat what had been laid down long ago. However there was much more to it than just that. The dangers to the freedom of science were not simple and fixed, but had fluctuated in form, extent, and national origin over the years, so that it was wise to restate ICSU's attitude to them. That this was no simple matter was brought out in a long debate at the General Assembly in Beijing in 1988.

I Fischer-Hjalmars referred to this issue when she presented her report on the work of SCFCS, describing the large number of cases it had dealt with: over 100, mostly about visas. The discussion then became very forceful, especially in the matter of repudiation. This is the device which has been used by some countries to interfere with the openness of the policy of free movement, not only of scientists. An applicant for an entry visas is asked to repudiate some policy of his or her own government which is dis-approved of by the government from which an entry visa is sought. This requirement for repudiation is repugnant to ICSU and to UNESCO. Japan used it for some time, perhaps because of its uneasy internal changes in accommodating to a new peace-time way of life. It has abandoned the practice. India was the last major country to give it up. But, while it was in being, argument about it could be fierce.

To take a few examples:[11]

The Nigerian representative (E U Emovon) drew attention to the difficulties arising in dealing with South Africa if ICSU Statute 5 were to be strictly interpreted. 'If it is decided that a nation cannot host an ICSU meeting because it would have to refuse entry visas to South African scientists or insist on repudiation clauses and thus discriminate against apartheid, this could also be seen as encouraging the present policies in South Africa.'

J E Fenstad (Norway) observed that Norway had a rule exempting participants attending ICSU meetings from certain visa requirements, but still had a policy of non-co-operation on cultural and academic affairs with South Africa. His advice was to deal with each situation 'in its entire complexity'. The representative of the Royal Society of New Zealand reminded the Assembly of the excellent example of the principle of universality in the presence at this meeting in Beijing of scientists from Taipei.

The Japanese representative regretted the loss to Japanese scientists because of the rejection by ICSU of an invitation to hold the present General Assembly in Tokyo on the ground that Japan insisted on a repudiation clause[12] in its visa application form. 'He reminded the Assembly that ICSU has no special competence in jurisprudence, and that the after-effects of decisions taken on these issues should also be considered.' The South African representative pointed out that his country is bound to play an important role in Africa as a whole and that it has a dedicated scientific community that took this African role seriously.

Other speakers added their own accounts of the conflict of principles which had emerged. The President (J C Kendrew) wound up by saying that the central issue of universality was so complex that the Executive might need the help of other professionals such as international lawyers and professors of moral philosophy, who might help in reformulating principles of universality which better allowed ICSU to deal with specific situations. This was done in 1989, and eventually resulted in the present text of the brochure *Universality of Science*, and the Statement in the *ICSU Year Book*

This discussion of the universality principle shows how ICSU had moved from including comfortable pieties in its outlook to seeking means of acting effectively in the legal field. The problems of SCFCS had never been easy. They had been made more difficult by having to take on the work of the Standing Committee on the Safeguard of the Pursuit of Science when this was dissolved (1986).

Freedom at home: SCSPS

There are many problems to be faced in freedom to visit another country for a short time for a limited time and purpose. What about freedom to gain and maintain a scientific livelihood in your own country? This is a very different matter. You can, for example, criticise a country for the way it behaves towards nationals of another country, and take steps to rectify or compensate for what you consider wrongdoing. But can you criticise it for the way it treats its own nationals? Can you find ways of acting from outside to relieve distress and obstruction inside? ICSU depends on goodwill which is always vulnerable, especially so if it seems to be contaminated by politics or to constitute an intrusion. Some members of ICSU thought there was an area in which ICSU could concern itself, and in 1974 ICSU created a Committee on the Migration of Scientists. The stimulus came from a long debate at the General Committee in Leningrad in September 1973. Interest centred on the results of a (non-ICSU) meeting of interested scientists held in Bellagio in March 1973 which had issued this statement:

> That the right to leave and return should not be denied to scientists and intellectuals as a class, particularly when a national government denies a scientist the opportunity to practise his profession for reasons other than the commission of 'non-political crime or acts contrary to the purposes and principles of the United Nations' [this being a quotation from the Declaration of Human Rights]. He should be free to move to another nation where he is welcome and where the opportunity is available to use his education or talent in creative scientific work.

This had been circulated by the National Academy of Sciences which had invited recipients to convey their views on it to ICSU.

Of several points put forward in the Leningrad debate, two illuminate fundamentally different attitudes, what one might call the western and the socialist, although neither adjective is adequate. The western looked at the scientist as an individual, who should be free to go where he liked as an individual. The socialist view (Ambartsumian, Kovda) was that, on the whole, the scientist owed his education, his opportunities, and his rewards to the state, to which he consequently owes his first duty. Moreover, since his activity as a scientist in a socialist country depends on the existence of state institutions, it is they and only they who should be considered to represent the science of that nation.

Eventually a Migration Committee was formed, but at the 1976 (Washington) General Assembly its name was changed to Standing Committee for the Safeguard of the Pursuit of Science (SCSPS). The committee touched on many sides of scientific work, like apparatus and materials.

In 1986 the Twenty-First General Assembly agreed (following a recommendation made to the General Committee in 1985) that its functions should be absorbed into those of SCFCS. In 1993 it was agreed that a new name was needed, but that the acronym was too familiar to change. So the name was changed to Standing Committee on Freedom in the Conduct of Science.

Measures of achievement

SCFCS with its new, wider mandate cannot intrude directly into the internal affairs of any country, but it can enquire and make representations. To be able to do this it needs to follow procedures first expressed at the Sixteenth General Assembly in 1976 'collect documents and analyze individual cases where *bona fide* scientists have been seriously restricted in the pursuit of scientific research or prevented from communicating with their fellow scientists'.

SCFCS and SCSPS both relied therefore on cumulative records, a body of archives (which were eventually united following their merger). The size of the archive can be inferred from the fact that the *index* entries alone to the correspondence on the Perth incident run to 5 pages. There are many hundreds of cases on file. It was not originally envisaged that the archive would expand to quite this extent and its present efficiency owes much to work done on its handling by Caspersson in 1973.

The archives are confidential, and even the terse Committee minutes cannot be reproduced in full, even if space allowed, but a few leading points reflect changes in the sources of the Committee's problems. We will take the decade 1984–93. The work went on continuously throughout each year, the Secretary responding to input from Unions, national academies, or individuals, and taking such action, on his own, with other members of the Committee or with ICSU members as seemed best. The number of cases reported to the annual plenary Committee meeting stayed fairly constant at around 40 per year, mostly being new cases, only a few being carry-overs from previous meetings. This is four times as much as Herlofson's original estimate in 1974. In each year some country or type of restraint is prominent. The same root causes keep on turning up, so a brief note of the topics may give an impression of the variety of causes of interference with free movement (mostly of individuals but sometimes of whole groups):

1984: antagonism to South Africa, difficulties of communication with USSR government bodies, the Arab-Israeli confrontation, the status of Taiwan;

1985: Canada on the wrong foot as unwilling host, instability of government in some countries of Latin America, Japan's repudiation requirement;

1986: (SCSPS cases are added from now on);

1987: status of China Academy of Science and Technology, interference with publication as a consequence of UN sanctions;

1988: problem of citizen of one country, resident in a second wishing to take up scientific work in a third, effect of commercial embargo rules on movement of publications;

1989: consequences of the Tiananmen Square incident, signs of softening of attitudes to South Africa;

1990: hunger strike by former ICSU President from the USSR, decision to avoid adopting a general view of Chinese problems but to take a case-to-case approach, India still hampering movement by repudiation requirement;

1991: China and the difficulty of distinguishing between UN Human Rights problems and freedom of science problems, Japan scraps repudiation requirement, how to find out when restraint on a scientist arises from a criminal offence rather than a civil/scientific attitude, new name for Committee.. consequences of the Eastern European turmoil;

1992: alleged discrimination on religious grounds found to be baseless and outside scope of committee, beginning of effects of conflict in former Yugoslavia;

1993: the first complaint of discrimination on the grounds of sex.

To repeat: any one of these types of case may turn up at any time.

The effectiveness of SCFCS (as is true of any useful committee) depends on personal involvement, The names of some of those worthy of mention have already been given, but others must be added here. From the ICSU Secretariat there were F W G (Mike) Baker, Executive Secretary of ICSU from 1969 to 1989, whose attendance at a number of meetings added greatly to the success of the deliberations. His successor, J Marton-Lefèvre, continues this excellent contribution. From the UK, D C Martin, Executive Secretary of the Royal Society was a member of SCFCS from 1963 to 1976. Such continuity is of the greatest value to growing effectiveness.

It would be easy to dismiss the aims of these committees as mere piety, but there is always a basis of realism underlying the ICSU debates on them. One quotation is enough, from the 1989 edition of the booklet *Universality of Science* which replaced the earlier *Advice to Organizers of International Scientific Meetings*: 'Only in an ideal world would all individual scientists be enabled to make the most fruitful use of their abilities. But a closer approach to this idea can be achieved by recognizing some of the problems and establishing generally accepted standards.'[13] And, always to be kept in mind: 'This concern is limited to scientists in their professional capacity'. ICSU was committed to staying outside politics, but as is evident in the brief list of SCFCS cases above, it is constantly alert to the professional consequences of political action. It is always open to it to put a non-political view to a national committee, which is then free to choose whether to assess its domestic political significance. As we have said in other connections, a national member of ICSU is not the central government of the nation but a body representing the whole scientific community of the nation. This sounds rigid, but in fact ICSU has more flexibility than if it were a governmental organisation. Since its national members are not nations but academies located within nations, it is not bound, unless it chooses to be so, by the restraints of the United Nations.

By the creation of SCFCS, of SCSPS, and by their fusion, ICSU has formed an instrument for maintaining the freedom of the scientist with greater scope than that of the UN itself. ICSU is not, of course, alone in being concerned for aspects of intellectual freedom: since 1945 there has been a whole spectrum of international conventions, some having agreed status as binding law in countries which have ratified them. Here again we find that hard good sense keeps the balance with emotion. The UNESCO Recommendation on the Status of Scientific Researchers (1974) refers to the need for scientists to respect public accountability. However, there is a difference between the imposition of rules and the arbitrary exercise of pressure. This difference can be recognised from outside a country, and measures taken to persuade a country that its behaviour within its own bounds has been contrary to opinion widely held outside. This may sound weak, but in the long run it works.

An afterthought: the work of SCFCS represents a general spirit of mutual aid in ICSU, which is most often shown in spontaneous acts of support. One can quote the example of an African scientist who arrived for an ICSU meeting in a European country but found he did not have the proper visa. His case was solved by the action of several experienced colleagues (not of his own country) with a speed (a matter of a few hours) which greatly surprised his government-official countrymen colleagues. One can tell of some scientists from one of the countries of the former Soviet Union who, attending an ICSU meeting in an unfamiliar country in the West, were robbed and injured on the street on a Saturday night. By Sunday their ICSU colleagues had seen to the repair of their wounds and had replaced their travel documents so that they could proceed easily on their way home. A Latin American scientist had a serious allergic reaction to a chemical in a pool in Brazil; his ICSU colleagues, acting with knowledge and efficiency, saved his life.

And so on: one could quote countless examples of such mutual concern. They are the local signs of something widespread which is summed up in the booklet *Universality of Science* (in the view of some, one of the most important achievements of ICSU). The insistence of ICSU and of individual Unions on the maintenance of its principles has had its effects not only on individual cases, but even on governments and their regulations. But the SCFCS members know that the price of freedom is unceasing vigilance.

Chapter 9

Living machinery: officers and staff

It would be misleading to draw a picture of the organisation of world science entirely in terms just of ICSU and its internal activities, but ICSU from the inside tells us a lot about science on the outside. In its early days one sees inside ICSU a reflection of what was most highly developed in science, its traditional physical and biological areas. Bit by bit the representation of these became subdivided, while new scientific fields also gained representation within the elected and appointed groups which grew within the ICSU body. These all had practical purposes, so that we can think of the management of ICSU as being a kind of living machinery.

General Assemblies

Any large organisation like ICSU must have a governing, deliberative organ at its heart. Action, inspiration, and continuity may be in the hands of voluntary, unpaid officers and permanent staff, but they cannot go on without scrutiny, renewal of mandate, or periodic discharge from financial responsibility. Such a deliberative organ does not usually have a continuous existence or continuous membership but is called into life from time to time, and made up of people who are summoned according to a constitution, in the case of ICSU its Statutes.

The ICSU General Assembly rises up, every two or three years, like a flower in the desert after rain, to debate, to pronounce, to criticise, to express satisfaction, to authorise, and, after a quite short time, to wither, leaving the dry leaves of a report, enfolding the seeds of change.

An ICSU General Assembly is a complex of personalities: individual and corporate. There are at least three kinds of corporate personality: the national scientific members, the scientific Union members, and the associates which have no voting rights. There are also at least three kinds of personal member: that is to say the members of the Executive Board (which includes the President and other elected officers), and, as observers, the representatives of partner-non-governmental and intergovernmental organisations, including UNESCO, all behaving in different ways. Exercising powerful influence, which has increased over the course of years, is the permanent staff, the Secretariat. Originally this was no more than some extra pairs of clerical hands for the Secretary and Treasurer. It is now the most important element in maintaining continuity of action and communication.

It was all rather simpler in the early days. At the last General Assembly of IRC, in Brussels, on 11 July 1931 (which was also the first of ICSU) there were present with voting powers: 4 members of the Executive Committee (Picard, Pelseneer, Volterra, and Lyons); 8 representatives of Unions (there were 2 additional officers of the IUChem (the later IUPAC), but they had no independent votes).

There were representatives of 23 of the 40 national organisation members, but this did not mean 23 actual persons; some acted in more than one capacity. Sir Richard Glazebrook, for example, represented and voted for South Africa, Australia, India, and New Zealand, as well as the International Union of Physics. General Georges Perrier represented and voted on behalf of Indo-China and Tunisia. Only one official observer from another body is noted in the official record: G Antonio Abetti, delegate for the Committee on Solar-Terrestrial Relations.

Altogether there were 42 persons present, exercising the votes of 8 Unions and of 23 countries, voting powers differing according to the nature of the resolution. The official business was concluded in one day, by lunch-time. The Second General Assembly of ICSU (1934, the first held under the Statutes agreed in 1931) had a similar composition, but lasted three days. In 1946 (the Fourth General Assembly) there were more than 60 people (the figure has again to be approximate because some delegates filled more than one role). In 1993 the Twenty-Fourth General Assembly in Santiago (Chile) consisted of the representatives of national members (each with one vote) and of the 20 International Unions (each with three votes except in matters concerning the dues of national members). This same General Assembly agreed to change this voting procedure to a new system (see chapter 16). There were present some 220 persons. The main proceedings lasted five days, but there were additional associated meetings, before and after. The attendance and the character of the Agenda had been quite transformed in the years since the General Assembly of 1934. There were now twenty Unions. For the first time since 1982 three new Unions were admitted.

There had been changes in the number and in the balance of influence of national members. The former British Empire representation had been exercised by proxy in 1931. At the end of 1931, the Statute of Westminster redefined the status of the British Dominions and formalised a *de facto* independence of what had become substantial nations. Australia had sent its own man to the 1934 meeting. Soon, with this change of the British Empire into a Commonwealth, not only had the 'old Dominions' their own scientists able to represent them; newly independent African nations were able to raise voices. Elsewhere other nations were emerging from scientific obscurity, so that in 1990 nations which had no recognisably independent scientific life in 1931, like Finland and the two Koreas, were represented.

Latvia had been a member of IRC and of ICSU in 1925; Estonia of ICSU in 1934. From 1940 they had lost their independence on annexation by the USSR and so did not appear at the 1946 General Assembly. The USSR did not become a member of

ICSU until 1955. Germany, in different guises, came and went and came again. It joined ICSU in 1931, and was replaced in 1952 by the Federal Republic of Germany (BRD). In 1961, the German Democratic Republic (DDR) also joined. After the events of 1990, they were replaced by a reunified Germany. (The membership at 1996 is shown in Appendix 2).

China had been a member since 1937, but in 1948 was represented by the Academy of Sciences in Taipei. In 1982 the China Association of Science and Technology, located in Beijing, CAST, also became a national member. It is one of the triumphs of good sense of the scientific community that both national members of China have continued to play a part in the activities of ICSU. There could be no better evidence of one of the leading principles of ICSU's non-governmental position than this: the national members of ICSU are national academies or associations, not governments. In 1982, CAST was admitted as a National Member of ICSU at the Cambridge General Assembly. Although the organisers were extremely careful to arrange the seating plan in such a way that China:Taipei and China:CAST would not sit next to each other but in separate rows, the arrival in the room of the younger delegates from Beijing was greeted with enthusiastic applause from all and warm embraces from their older colleagues (and, in some cases, teachers) from Taipei. This was the culmination of years of negotiation and a very moving moment in ICSU's history. In 1988 the ICSU General Assembly was held in Beijing and a delegation from Taipei attended, the first official delegation of its kind.

The hiatus of the 1939 war underlined a significant factor in the life of ICSU (as of any such body), namely, that a small number of interested, devoted, people can keep a body alive for a long time, while the authority of a governing assembly is not available. Between the General Assemblies of 1939 and 1945 ICSU stayed alive in the hands of a London-based Secretariat and individual officers of Unions in many countries, belligerent and neutral. It was thus possible to revive the General Assembly, in form and authority, very quickly – so quickly, in fact, that it was ready to authorise negotiations with the new UNESCO immediately UNESCO was in a position to assess its own scientific character.[1]

From 1945 onwards the General Assembly was aware that, although ICSU had always rightly felt it was acting on a world stage, the coming of the United Nations in place of the League of Nations, and of UNESCO in place of IIIC, had changed the scenery of the stage.

One must not think that ICSU meekly sought the patronage of the new body. It was the other way round, as one sees from the memoirs of the first Director-General of UNESCO, Julian Huxley:

> It had been laid down by the London Conference [setting up UNESCO] that UNESCO could seek help in technical matters from non-governmental International Agencies concerned with subjects within UNESCO's purview; and, if necessary, aid in the creation of new ones. It was on Needham's advice that ICSU, the International Council of Scientific Unions, became the first of such bodies to be attached to UNESCO. We provided it with rooms in UNESCO's headquarters in Paris, and salaries for its staff:

previously the Cambridge professor who was its secretary – and sole executive – had to dictate all its correspondence in his College rooms!

The ICSU gave UNESCO much valuable advice – on the peaceful uses of atomic energy, on regional centres for scientific co-operation and exchange of scientific knowledge, on the calling of international scientific congresses, and on liaison with other International Agencies concerned with science, such as FAO for agricultural science and applied ecology, and WHO for medicine, physiology and social well-being.[2]

The liaison with other international bodies indicated by the above quotation developed considerably as a concern of successive General Assemblies.

The changes in scale and in scope entailed changes in the way the General Assemblies meetings were set up and conducted. The official record of the plenary sessions of the General Assembly of 1952 reads rather strangely nowadays, as if it were a continuous verbatim record of a long conversation between a large group of friends. It is quite difficult to work out from this report alone what the Agenda could have looked like. The Finance Reports are separate and more distinctive; the discussion is well focussed, as so often happens when a group of people of different characters get down to their common interest in money. The separate reports from Unions and Commissions, which were circulated in print first, are generally clear and well sectionalised. But to identify decision-making one has to read every word of the main report.

Presentation then changed, and the report of the Seventh General Assembly (1955) is to be found in a new publication: the *Year Book of ICSU* which had been introduced in 1954. This report of the 1955 debates was edited with an eye to easy retrieval of information.

Venues

There are many ways of looking at the history of a multi-faceted organisation. Even a simple list of the venues for the General Assemblies tell us something about the extent to which the leaders of ICSU were justified in speaking of it as a world organisation. One has to be careful not to jump too quickly to conclusions, but the sequence tells its own story.

The IRC's General Assemblies (1919, 1922, 1925, 1928, 1931) had all been held in Brussels. The 1931 General Assembly became the first General Assembly of ICSU (July 1931). The order that followed was:

1	Brussels	July	1931
2	Brussels	July	1934
3	London	May	1937
4	London	July	1946
5	Copenhagen	October	1949
6	Amsterdam	October	1952
7	Oslo	August	1955
8	Washington DC	October	1958
9	London	September	1961
10	Vienna	November	1963

11	Bombay	January	1966
12	Paris	September	1968
13	Madrid	September	1970
14	Helsinki	September	1972
15	Istanbul	September	1974
16	Washington DC	October	1976
17	Athens	September	1978
18	Amsterdam	September	1980
19	Cambridge (UK)	September	1982
20	Ottawa	September	1984
21	Berne	September	1986
22	Beijing	September	1988
23	Sofia	October	1990
24	Santiago	October	1993
25	Washington DC	September	1996

The five of IRC and the first six of ICSU were in Western Europe, and it would be easy to take a sentimental view of history and suppose that this was because IRC and ICSU had their strength in the countries which laid the foundations of modern science. In those countries had been founded the great academies which were able to provide some degree of continuity to any large scale scientific enterprise. However, this view will not do. Science was not the special perquisite of the Atlantic countries. Russia was becoming increasingly important in the 19th century. Before Bismarck, the German states were as welcoming to the scholar as any country to their west. After Bismarck, the German academies were able to take that lead in seeking international collaboration which led to the formation of IAA. Or one can take a cynical political view and suggest that the General Assemblies of IRC from 1919 to 1931 were held where lay the strength of the Allied victors in the 1914 war.

Both these considerations, cultural and political, no doubt played a part in the decision about where to hold any General Assembly, but ICSU could not decide of itself where to go. There had to be an invitation from some country which could afford it. Entertaining an important international body, even one which has funds to pay some of the overall expenses, costs a great deal, some of which must be borne by the host. For the first few years of its life ICSU had to hope for an invitation from one of its early strong members. So one sees a progression from Brussels, which was a key city because of its status as the legal domicile, to London, which was the home of one of the most influential of the national academies, to Copenhagen, to Amsterdam, and then to Oslo, all of them now grateful for freedom from an occupation they could not forget.

The United States involvement in ICSU was of long standing. (Hale had been as creative as anyone in its formative days.) A meeting of the Executive Committee of IUGG had been in session in Washington at the outbreak of the 1939 war. It was natural that the first move of a General Assembly to a venue outside Europe should be to the domicile of the National Academy of Sciences, Washington DC. The Atlantic was rapidly getting narrower: in time, that is to say. It had been possible, but very unusual, to fly the Atlantic in 1939. By 1958 it was commercial commonplace. From then on there was no geographical limit to the siting of a meeting, only the limits of finance and the exigencies of policy.

Back to London in 1961, and then in 1963 the consolidation of that contact with the community of the German-speaking academies that had been so significant at the beginning of the century: a General Assembly in Vienna. The meeting of the Tenth General Assembly in Vienna, 22–29 November 1963, rearranged the structure of ICSU, but this is hardly likely to be remembered as vividly as the event of Friday 22nd November, the assassination of President Kennedy, news of which reached the members while they were attending a reception by the Austrian Academy of Sciences.

Another death was to cast a cloud over recollections of the next General Assembly, although it did not occur until after the Assembly was over and delegates were getting back to their professional activities. On the instance of H J Bhabha[3] the distinguished Indian physicist, a General Assembly was held for the first time outside the western world: in Bombay in 1966. It may be that being in India helped the delegates to take a more detached view of ICSU's problems than they could have done in any European setting. His action was deeply appreciated by all. A few days after the General Assembly he returned to Europe to die in an air crash on Mont Blanc on his way to a meeting with fellow physicists on the peaceful uses of atomic energy.

This was a great loss to international science, but the Bombay General Assembly left a lasting legacy. It promoted ICSU's growing interest in science in the Third World, without which it could never be a true world organisation. At that Assembly COSTED was set up.

The Paris meeting of 1968 took place in a tumultuous year. The Paris student riots and the Warsaw Pact intervention in Czechoslovakia gave a tension to many a conference held that year, including the IUHPS Congress which also took place in Paris. September 1970 found the General Assembly being addressed by Spanish official hosts anxious to look to a future which they knew already was bound to be one of radical change with the passing of Franco, whenever it should come (and it could not be far off).

At the Helsinki meeting in 1972 some tidying up of the Statutes was agreed: the Executive became an Executive Board, the idea of a President-elect was dropped. These changes may look like merely playing with terminology, but occasional adjustments had to be made to keep the working machinery of ICSU in tune with expanding membership. Sometimes, however, the examination had to be more searching. In 1985 (at the Ringberg Conference, which needs a separate chapter) ICSU took a good look at itself and its organisation, at its relations with other bodies, its own and their joint purposes.

Istanbul as the venue for the Fifteenth General Assembly (1974) marks another broadening of ICSU scope: Turkey, lying between Europe and Asia, emphasising the change in world conditions which was once fearfully described as 'the decline of the West' but now has to be seen more optimistically as the equalising of development opportunity. The undercurrents of the spread of science are illustrated by a conversation between one delegate and a student acting as a guide. The delegate complimented the student on his excellent English and asked him whether he had learnt it in England or America. The student replied that he had never been out of Turkey, but since the range of texts he needed was available only in English he had no hope of advancement unless his English was equal to reading anything he might come across, which demonstrated the importance of the language factor in scientific communication.

For some years the General Assembly remained in the West. To go to Washington in 1976, the bicentenary of United States Independence, was both obvious and courteous. Athens (1978), Amsterdam (1980), Cambridge (1982), Ottawa (1984), and Berne (1986) were all comfortable choices.

Then came a bold adventure. China had always been a problem. To hold a General Assembly in Beijing (1988) seemed to turn one's back on the past, and look to a future of co-operation over large physical territories and with a large number of able scientists. To a large extent this has been achieved, notwithstanding the suppression of a mass demonstration in 1989, seen all over the world on television.

The venues of the next two General Assemblies could hardly lie further apart, geographically or culturally: Sofia (1990) and Santiago de Chile (1993).

Governance

The form of the General Assembly has remained much the same, reflecting national scope, and consolidating the continuous work that goes on in between times. That continuing work has been done by committees and *ad hoc* groups, some stemming from the Statutes, some specially created. Some exist just to serve the General Assembly, some have acquired a permanent existence.

From 1931 to 1946 ICSU managed with the simplest of structures. The General Assembly elected a President, two Vice-Presidents and a Secretary-General. The resolutions of the General Assembly were put into effect by an Executive Committee, consisting of the President, the two Vice-Presidents, the Secretary-General and two delegates from each of the Unions, these delegates being nominated by their Unions. There was no Treasurer yet. The funds of the Council were managed by the Secretary-General, who was also responsible for all Council correspondence. The Executive prepared an annual report for all Council members.

At the Fourth General Assembly (1946) the Executive Committee's proposals for some changes were accepted, simple enough, but the beginning of a recognition that times and needs had changed. It was agreed that the retiring President should be included in the Executive Committee. In the report of the General Assembly now occurs the word 'Bureau' (which continued in informal use for a considerable time) although the adoption of such a term is nowhere included in the report of the discussion. It was formalised from 1949 and later abandoned.

The Fifth General Assembly (1949) spent a long time on a proposed revision of the Statutes which had been prepared by a committee of the Executive under Professor

Kruyt, and the final form presents the structure of which the backbone remains to the present.

In 1963 the General Assembly accepted recommendations of the Future Structure Committee, chief amongst which were:

(a) abolition of the Bureau and Executive Board, and their replacement by an Executive Committee (President, Secretary-General, Treasurer, Past President, fourteen Union representatives and ten national representatives);

(b) establishment of a category of non-subscribing, non-voting National Associates;

(c) abolition of the General/Specialised distinction between Scientific Unions.

(Of these three (a) was superseded by later structural changes. By 1973 the main committee to meet between General Assemblies was a General Committee.)

The Statutes do provide for committees for special activities, on nomination by the Executive. When the provision was first included it read as if such functions were expected to be modest in size. Some came to be very large indeed, and one of the major functions of the Council (as the tale of IGY, WCRP, and IGBP will show). In the first five years of its post-war activities, ICSU developed, on its own or in collaboration with Unions, a large number of committees and Mixed Commissions to address specific scientific problems which did not fall distinctly within the province of some particular Union.

In every year between General Assemblies – and at every General Assembly – the so-called General Committee met to screen newly proposed scientific initiatives and to set scientific priorities. It also had the task of reviewing periodically the work of the Special and Scientific Committees. It consisted of the Executive Board, representatives of all Unions and a limited number of national members. After a while, meetings of this Committee tended to develop into mini-general-assemblies as matters of finance (not within its mandate) appeared on its agenda. This was corrected in the early 1990s and the General Committee reverted to its original mandate.

Over the years several types of committee developed:

– those serving the General Assembly only, such as the Nominations Committee;

– those giving assistance and advice to the Executive Board and or the General Assembly on a more permanent basis like the standing Finance Committee, the SCFCS, the SCMSS, ACE, and the ICSU Press Service;

– most importantly, and of an altogether different calibre, the special and scientific committees created for interdisciplinary international work and for action in areas of common concern. These creations constituted, of course, ICSU's main *raison d'être*.

Each of these committees has its own history. Let us take a look at some of them.

What organisation might be a member was always a matter for discussion, as we find in the earliest reports of General Assemblies, but ICSU shares this characteristic with many clubs: a candidate for membership must expect to undergo some sort of scrutiny and elective procedure. Up to a point ICSU could carry out this scrutiny in the General Assembly itself (which means, as always, that contentious matters were probably resolved in the corridors). This is workable up to the point at which the membership becomes too big for easy discussion, and membership of the organisation becomes an aim of the ambitious. So in 1966 ICSU set up an Admissions Committee to put proposals to the General Assembly.

The records of the discussions in this Committee and in the General Assemblies show that application for membership of ICSU was not always immediately successful. No application was accepted without discussion, many were deferred for investigation, and a few were rejected.

The developing structure of ICSU is a reflection of the developing structure of world science. You may say that science just grows, aimlessly, and that any pattern put on it by the historian or biographer is artificial. But few scientific advances, other than those which stem from pure mathematics, are the product of lone workers. Communication is of increasing importance, and ICSU has been progressively more concerned with means of facilitating communication. One of the best contributions it can make to this is to improve and advance its own machinery. This it had done as time went by, but it had left it to two separate groups to advise, one a group which discussed structure, and one which, from time to time, looked at the current relevance and effectiveness of the Statutes. In 1980 the General Assembly agreed that these were two sides of one coin, and that henceforth there should be one *Ad Hoc* Group to consider Structure and Statutes together. Finally, in 1993, this Committee was merged with the Admissions Committee into one Standing Committee on Membership, Structure, and Statutes.

To some it might appear surprising to find that ICSU managed to survive for years without one committee which anyone setting up a new organisation might consider essential, namely a formally constituted Finance Committee. Up to 1939 ICSU's finances had been managed first by a Secretary and then by a Treasurer, with some professional advice from auditors. In 1946 reference is made to a Finance Commission. Thereafter a report is received from a Finance Committee. Some of the subsequent reports from Treasurers give an unflattering picture of the ability of some ICSU members to keep their houses in order. It took several years for some Unions to pay dues. The most embarrassing thing for ICSU was the slowness with which some Unions prepared accounts of the use of UNESCO subventions, which put ICSU in an undeserved bad light. The cause may have been indolence on the part of Union Treasurers, but a more likely factor is that some Unions were dependent on the goodwill of professional scientists who had to face money matters of some novelty and difficulty. More than one scientist elected Treasurer of his Union, quite capable of dealing with financial matters expressed solely in the currency of his own country, found it was a very different matter dealing with accounts in four or five hard currencies and the needs of member countries who had none of them.

Committees can arise from slow or from sudden change, and can change form. The Ethics Committee came into existence because of the gradual appreciation by ICSU members of the growing public interest in, and the growing responsibility for, study of the ethical questions which were encountered more and more often in scientific work. The ethical aspects of environmental questions are obvious. So are many ethical medical or biological questions. An Ethics Committee was formed on the instance of

IUHPS to bring order into what might have become uselessly random discussion, but it became clear that the academic philosophers were pointing the way to a commitment to responsible action. In 1990 the Executive Board therefore decided that it must itself be responsible for the review of ethical considerations in science on behalf of the whole ICSU family, relying for advice on IUHPS.

The transformation of Eastern Europe in and after 1990 brought about social and economic changes which affected the practice of science at all levels, not only within Eastern Europe, but in the world at large. All manner of organisations had to reconsider membership, and publication. The scientific community and the members of ICSU in particular could not just observe: they had to accept responsibility towards colleagues from whom political differences had never been able to divide them. An *Ad Hoc* ICSU Committee was set up in 1990 to consider the problems of scientific activity and organisation arising, and likely to arise, out of a new situation, and to explore ways in which ICSU might help to relieve this problem. It became a Special Committee in 1993 (COMSCEE).

There are thus many Committees formed to promote the internal working of the ICSU machinery without very specific scientific subject content. Other kinds of body created by ICSU are clearly more scientific in nature and have a more external aspect, but all may be represented at the General Assembly in one capacity or other.

As time went on, the ICSU family became an international network linked by phone, by fax, and by electronic mail, which has enabled ICSU to obtain quick answers from its members and from its committees on a wide range of questions. Often a specific group of members can be asked to propose names of candidate members for some new expert committee. Each of these must have, as far as possible, a composition balanced as to required disciplines, geographic origin and, more recently, as to gender. This generally works well, but tales are told that confusion has been known to arise when two prominent scientists in one country had the same name and the 'wrong' person was asked to serve instead of the one envisaged. In one case the 'wrong' person accepted and turned out to be an excellent committee member!

Staff

The change of administrative style marked a change in the character of ICSU, one which came about because of ICSU's leading role in the organisation of the International Geophysical Year. It is interesting here to hark back to the reasons for the replacement of IRC by ICSU. The aim of IRC to stimulate research work had been considered inappropriate, and ICSU had replaced it, leaving the initiation of research to the Unions where that research fell clearly within the scientific scope of one particular Union. However, ICSU was now able to do what IRC could not. It provided a machinery for co-ordinating efforts which, if left to individual Unions, could not (or not so well) be brought together. The Second International Polar Year had made this clear. Although IRC and ICSU had not been prime movers in creating this 1932 IPY, the way in which it had been organised offered a clear indication of the way large enterprises might best be set up in future. The eventual success of the IGY organisation justified this confidence.

It was not an overnight change: more of an evolution. Relations with UNESCO had been so manifestly important that as soon as a satisfactory accord had been agreed in 1947 an ICSU–UNESCO Liaison Officer had been appointed (the 'staff' referred to in Julian Huxley's reminiscence quoted above). The first was A Establier (who had represented IIIC at ICSU meetings).[4] He was succeeded in 1949 by Ronald Fraser,[5] who built up relations between ICSU and other international bodies. In spite of the value to ICSU of the close contact with UNESCO, it was felt that the Secretariat should be in closer touch with the Secretary-General (at that time A V Hill).[6] In the winter months of 1953–4 the Secretariat moved to London (the Administrative Secretary now had his office at 29 Tavistock Square, London; the Secretary-General, A V Hill received correspondence at the Royal Society, then at Burlington House, Piccadilly, London).[7]

The Secretariat was no longer to be considered as just the tool of the Secretary-General, but as the main element in the permanent structure of ICSU. It could therefore operate from an independent base, and the first such base was provided in 1958 by the Royal Dutch Academy of Sciences at the Palais Noordeinde in the Hague. In 1961 Ronald Fraser vacated his post (to become editor of the short-lived *ICSU Review*). He was succeeded by André Decae, whose post was renamed Administrative Secretary. An Accountant, Alistair McLennan was also appointed.

Being host to ICSU now carried some prestige. When ICSU had to move because the rooms in the Noordeinde Palace at The Hague were to be refurbished, there were several attractive offers. The Assembly accepted the offer from the Consiglio Nazionale delle Ricerche of rooms in Rome. The staff expanded again, to something like its present number, first by secondment of Italian staff, and later by the transfer of staff from FAGS, IGY, and IBP. Staff came and went too often for comfort or efficiency, and all looked forward to the day when the staff could settle down to some degree of stability. McLennan left in 1963, A M Rao being then appointed as Accountant. In 1965 Decae and Rao left, to be replaced by F W G (Mike) Baker as Executive Secretary and G Bollenbach as Accountant.

This was the point at which the Secretariat of ICSU took on a long-lasting stable form. When Baker retired in 1989, he had served ICSU in one capacity or another for 32 years, and so had provided an element of personal continuity which far exceeded that of any other officer, honorary or staff.

ICSU moved house again in 1972. It seemed expedient to move to Paris, if possible, because of the growing connection with UNESCO which had been firmly established there since its foundation. There were long negotiations with the French government, which ended harmoniously with ICSU accepting the pressing invitation of the French Académie des Sciences to move to the Hôtel de Noailles in Paris. ICSU has stayed much longer at this address than at any other.

Bollenbach retired in 1978. The opportunity was taken to recruit another strong personality: Julia Marton-Lefèvre. Titles of posts reflect authority. Julia Marton-Lefèvre's post was first called Assistant Executive Secretary, then Deputy Executive Secretary. On Baker's retirement she became Executive Secretary. In 1991 the Executive Board renamed this post Executive Director. This has happened in many organisations: a post which could be held originally by a modest clerk evolves over the course of years and through many occupants into a key position. This can only happen when there is a history of devoted service by a succession of persons of superior quality. ICSU has been fortunate to know so many.

There was a period (1975–6) when it was thought that a new type of appointment was needed which might be called *Scientific Director*. The idea was that, since ICSU dealt continuously with scientific matters at an advanced level, its administration should be in the hands of a scientist of the highest attainment and repute. The suggestion eventually foundered on three obstacles: too big, too expensive, and too dominant. It was thought that with such a person in charge the Secretariat would need to be expanded to cope with the extra work-load such an arrangement would generate inside the system. Better, it was concluded, to spread the scientific load over the Unions and special committees. It would also be costly: a Scientific Director could only be paid at the highest rate, and there would be the additional cost of the new kind of support staff. It would also be in danger of having far too much scientific policy-making in the hands of one person who, after all, could be a good scientist in one discipline only. Serious discussions were held with the Ford Foundation about the funding of an administrative trial, but it was decided by 1976 that such a step would be unwise and that the progressive improvement of the existing administrative machinery would be a better way forward.[8]

However, in the early 1990s the idea took shape in an entirely different form through the appointment of a staff officer at the Secretariat to assist ICSU in its work in Environment and Earth System Research. Such officers working part-time for a few years were first-rate scientists seconded, or paid for, by one or more national scientific members.

Under the first Executive Director the staff of eight worked as a coherent team. Although each had specific duties, all of them were kept informed of what went on so that most could, to a large extent, stand in for a colleague. In 1993 there were an Assistant Executive Director, a financial officer, and four administrative assistants. All used word processors, and the administration and communications was computerised as far as modern technology permitted. A library keeping constantly updated information on ICSU Members and Committees was set up and the archives were catalogued so as to facilitate access.

The size of the ICSU secretariat was, and remains, quite small compared with the administrative machinery of most of its members. Certainly all national members have larger staffs and so do several of the Unions, IUPAC being a case in point. In turn, some national member associations of IUPAC have larger offices than their parent, e.g. the American Chemical Society. It should also be realised that the UN organisations engaged in partnerships with ICSU have more staff and resources, so that, in co-operative efforts with bodies outside as well as inside ICSU, the Paris ICSU secretariat sometimes finds itself forced into peaks of intensive activity to keep pace with the work. Most of the time however, the small size and quality of the secretariat enable ICSU to set a pace which can only slowly be followed by the larger and more bureaucratic administrations.

Creations: Special and Scientific Committees

It is easy to mock committee-making, but the meeting of like minds is one of the most powerful of creative influences. A committee can be set up to puzzle something out. It can be set up to manage a programme. One of the most important features in the history of ICSU is the way it developed its creative power by bringing together people who recognised common aims and devised means of achieving them. This is true of other organisations, but a distinctive feature of ICSU is that it subsists as much on the essentially creative character of scientific investigation as it does on human personality. Any account of the works of ICSU contains many acronyms, most of them signifying one of two things. One acronym may stand for an organisation which grew out of the coming together of people who have recognised a common aim and who initiate action; another may stand for a field of currently productive new scientific activity which draws in people who feel they have something to contribute.

Special Committees are set up when a limited lifetime is envisaged. Some are eventually abolished, others became Scientific Committees when their *raison d'être* has been proven.

We see here another essential difference between IRC and ICSU. IRC was a single group which hoped to operate a single mechanism for the development of new work. ICSU relies on no one system, but works out the best way to deal with new needs, whether they arise from discussions within the ICSU structure itself or from Unions, or from any other body with which ICSU has an association.

The Report of the 1949 General Assembly contains the forerunner of what was to be an important publication. It had been agreed that ICSU should publish a brochure, for general distribution, containing a detailed account of the Council and its activities. Pages 109 to 197 of this 1949 Report consist of a first draft of this brochure, of which a revised and updated version appeared eventually as the First Edition of the *ICSU Year Book* (first published in 1954). In this we see a complex of ICSU Committees, Unions and their internal commissions, with a few mixed commissions dealing with limited areas of interest. Very soon, however, new larger bodies began to appear as substantial enterprises required their creation.

Here are a few with dates of creation. Some survive, some have changed, some have disappeared (see Appendix).

   Year of creation

1956 FAGS Federation of Astronomical and Geophysical Data Analysis Services
1957 SCOR Scientific Committee on Oceanic Research

1958 COSPAR Committee on Space Research
1958 SCAR Scientific Committee on Antarctic Research
1964 SC-IBP Scientific Committee for the International Biological Programme
1964 COWAR Committee on Water Research
1966 COSTED Committee on S&T in Developing Countries
1966 SCOSTEP Scientific Committee on Solar-Terrestrial Physics
1966 CODATA Committee on Data for S&T
1967 GARP Global Atmospheric Research Programme
1968 CTS Committee on Teaching of Science
1968 Panel on World Data Centres
1969 SCOPE Scientific Committee on Problems of the Environment
1976 COGENE Scientific Committee on Genetic Experimentation
1979 CASAFA Scientific Committee on the Application of Science to Agriculture, Forestry, and Aquaculture
1980 WCRP (together with WMO and later IOC of UNESCO)
1986 SC-IGBP Scientific Committee for the International Geosphere-Biosphere Programme
1986 COBIOTECH Scientific Committee for Biotechnology
1990 SC-IDNDR Special Committee for the International Decade for Natural Disaster Reduction
1992-4 The Global Observing Systems (GCOS, GOOS, GTOS with UN bodies)

All of them must submit annual reports to the Secretary-General and, as mentioned above, the General Committee periodically (once every five years or so) reviews the performance of each body. When appropriate the General Committee may recommend the modification or abolition of a committee, which does happen regularly. There were only four related bodies in 1960. In 1982 the President, Professor D A Bekoe, said that ICSU united 39 international scientific and technical organisations compared with only 20 ten years previously. Some of these were ICSU creations, 20 were Unions, some had taken Associate status. By 1993 the number of bodies listed in the Year Book as having one kind or another of working relation with ICSU had risen to some hundreds. The variety may seem very wide, but its size is enough to enable us to identify three main reasons for their creation

  – the need to foster new areas of science not pertaining to the mandate of one single Union (e.g. SCOR, COSPAR, COGENE, COBIOTECH);
  – to promote science in all parts of the world (e.g. COSTED, IBN, COMSCEE);
  – the need to address major world problems of an interdisciplinary nature by assessment (e.g. SCOPE) and by co-ordinated research (e.g. IGBP) sometimes as partner with one or more UN bodies (e.g. WCRP).

The history of rocketry theory is much older than the events of the thirties and forties would suggest,[9] but the post-war experience in IGY, that scientific observation could be carried on systematically outside the bounds of the earth system, suggested that ICSU could usefully keep a wide view of activities in this field. A Committee on Space Research was set up at the Eighth General Assembly meeting in Washington in 1958 when the IGY programme was being wound down. This COSPAR committee was concerned more and more with massive expensive structures, entailing considerable financial outlay.

It is easy enough to see why concern should be felt in respect of the space enterprises, which involved massive engineering projects. Moreover, it was to be influenced, in due course, by political considerations to a greater extent than any other ICSU body.

Another committee, however, concerned experimentation at the other extreme of magnitude: within the cell and within the molecular structures responsible for heredity. The elucidation of the role of DNA and then the elucidation of its structure were Nobel Prize achievements[10] which sent ripples of excitement over the whole community of science. The variety of studies opened up was very great, in the study both of the normal and the abnormal. So COGENE (Scientific Committee on Genetic Experimentation) was set up to act as another of ICSU's unifying tools. There were, as with all such advances, ethical issues, in the consideration of which COGENE could offer guidance, such as the consequences of any moves from the 'pure' to the 'applied and technological' and then to the commercial.

In its simplest terms, while traditional methods of producing new kinds of creature by the age-old methods of the farmer and stock-breeder or the pigeon fancier, intervention in the fundamental machinery of succession and inheritance raised problems of a science-fiction type seriousness. Genetic experimentation had to be watched so as to keep a balance between the benefits of widespread use of beneficial discovery and the twin dangers of biology out of control and of irresponsible commercial exploitation.

When, in 1966, the General Assembly met in Bombay, away from the limiting atmosphere of the European-Atlantic world, under the Presidency of the far-sighted Harold Thompson, it was time for ICSU to reassess its world responsibility and to make certain that the role of science in the future prospects of the developing countries was firmly fitted into the ICSU machine. The proposal to set up the Committee for Science and Technology in Developing Countries (COSTED) did not have a clear run. There was some uneasy discussion with the representative of UNESCO, Professor A Matveyev. He thought UNESCO, which had its own approach to 'development', would want to see evidence of any utility the new committee might ultimately have alongside UNESCO's own efforts in this field. It was suggested against this that UNESCO's own work in science in developing countries was shallow and ineffective. A common view was arrived at, and COSTED came into existence (with Professor P M S Blackett as first Chairman), to begin a life which continued until 1993 when its structure was adapted slightly in order to combine with ICSU/UNESCO International Biosciences Networks. The striking words in its title are 'Developing Countries', but of equally great significance is the conjunction of 'Science and Technology'. The same two words appear in the title of another Committee, CODATA (Committee on Data for Science and Technology), also set up at Bombay in 1966. Without fuss or argument ICSU had recognised that science and technology could no longer be thought of as separate realms of thought or activity. The *International Critical Tables for Science and Technology* had long been recognised as of continuing value in many areas of science and technology, but this title did not represent any systematic view of endeavour. A new overview was needed, and this ICSU now provided, recognising in the title of the Committee that it stood for a service common to pure and applied science.

As time went on more of the work of ICSU was generated by the coalescence of problems which encompassed the whole globe. Humankind lives on the solid substance of the earth, but ranges over a continuum which is to a great extent water. In 1957 ICSU seems to have heard the voice of the poet crying 'Behold the sea, itself', and set up a Committee to further international scientific activity in all branches of oceanic research (Scientific Committee for Oceanic Research: SCOR). No ICSU Committee has developed so multifarious a national membership, working-group structure, or affiliated association membership. Considering the variety of oceanic phenomena, physical and biological, this is hardly surprising.

The fascination of the sea is as old and widespread as poetry and science. Now we are conscious that we live, breath by breath, in a thin layer of air between solid earth and the hostile vacuum of space. Anxiety about the environment is an old story: as soon as coal began to be used in large quantities there were protests about pollution. The growth of the chemical industry in the 19th century produced much protest and consequent legislation. These were often local problems, attacked by local or national legislation. In the lifetime of the compilers of these pages, pollution problems have become acute and grave, but only very recently have they been looked at internationally. ICSU was relatively early in the field of action, setting up its Scientific Committee on Problems of the Environment (SCOPE) in 1969 in parallel with the preparation process for the UN Conference on the Human Environment (Stockholm, 1972).

These Committees are examples of ICSU creations. ICSU works with and through other bodies with which it can set up close working relations. The 'living machinery' of ICSU is not limited to what it can generate inside itself. As in the case of WCRP, GCOS, GOOS, and GTOS, ICSU is a partner on equal footing with one or more UN organisations. The rationale for this is that ICSU has the reputation for being unbiased and objective, and can ensure the scientific vitality and quality of programmes by bringing in the best scientists. These joint programmes together with IGBP, a 100% ICSU creation, constitute together the largest and most complex research and observation programme ever undertaken with respect to the Earth System.

Co-operating organisations

If we vary our image of ICSU sometimes as an organism, sometimes as a network, and sometimes as a family it is because none of these images is quite adequate. Let us try setting ICSU in a 'community'. The community of scientific bodies with which ICSU is associated is very large, although, if one excludes intergovernmental organisations, none looks quite so comprehensive. There are several kinds of connection provided for in the ICSU Statutes. A body may adhere as a National Scientific Associate, as an International Scientific Associate, or as a Regional Scientific Associate.

Some are older than ICSU, like the Pacific Science Association, created in 1920, which adhered to ICSU in 1970 as a Regional Associate. Of the Scientific Associates the oldest is IUFRO (International Union of Forestry Research Organisations) founded in 1892, reorganised in 1971, adhering to ICSU in 1980. Nearly the oldest is one which is perhaps more widely known than most: FID, the International Federation for Information and Documentation. Founded in 1895, adhering to ICSU in 1970, it has had a widespread influence in the development of the fundamentals of classification and other information techniques which are common to all learned pursuits. At the time of the General Assembly in 1993, 10 National Scientific Associates, 4 Regional Scientific Associates, and 23 International Scientific Associates. Of special importance are TWAS (1985) and IIASA (1987).

Naturally, ICSU has several levels of association with the principal intergovernmental bodies, the most important ones being UNESCO (1946) and WMO (1960). Other examples are the International Atomic Energy Agency and an organisation which is concerned with practical issues: the International Telecommunications Union (arranged in 1962). Specialised consultative status with the Food and Agriculture Organisation was granted in 1963, since when there have developed many collaborative groups such as joint committees involving ICSU and FAO with the International Biological Programme and SCOR.

ICSU has, furthermore, in recent years acquired consultative status with various UN bodies involved in environmental issues like the UN Commission for Sustainable Development, the Intergovernmental Panel on Climate Change, and the governing bodies of various conventions. Representatives of ICSU regularly exercise their right to speak in meetings of these bodies as it has done already for many years in the General Conference of UNESCO and the governing body of WMO.

Lists of acronyms of international this-and-that convey no sense of personal activity, of any productive intervention in the human or natural worlds, but that is what all these bodies were doing. It is rather easy to see the affairs of the world in terms of the political news of parliaments and ministers or the condition of currency or the latest war. Important they are, but at the same time thousands of men and women are involved in the promotion of beneficial change through the ABCs and XYZs we see in the ICSU Year Book.

Although ICSU is insistently non-governmental, there are many occasions in which persons who have had high-level administrative appointments in the service of their governments have also been leading scientists with a concern for ICSU. ICSU officers have, for example, occupied the posts of President, Foreign Minister, or Minister of Science in their countries. Whenever these persons attended ICSU meetings, however, they left their governmental brief-cases and, in some cases, bodyguards, safely outside the meeting rooms and then continued to contribute to ICSU business as independent scientists. (The ICSU staff, never comfortable with inaction, have got used to inviting the bodyguards to help with chores round the Secretariat: setting table for luncheons in the ICSU garden, photocopying, or even washing dishes.)

In this respect it is worth noting that, during the entire Cold War period the

Executive Board always had at least one scientist from a Warsaw Pact country (usually the USSR) and one from the USA which avoided any semblance of political bias. This was not a statutory obligation, but an unwritten arrangement accepted by all for the good sense it made.

'Living machinery' is not too fanciful a phrase for the large numbers of people involved in ICSU's activities, but it fails in one respect if the word 'machine' suggests something which stays the same. ICSU always has the means to change what it does and how it acts by resolution of the General Assembly, but in such a meeting many things have to be done in a few days. ICSU (like any other such body) sometimes needs a longer and more searching debate than can be provided for in a statutory way. Two such long debates took place in 1985 and 1990. The issues that surfaced tell us something about ICSU's past and what seemed at that time to be its future. We shall use these Ringberg and Visegrad meetings to help us see the living machinery of ICSU in its world setting.

Chapter 10

Growth: within science and outwards

Science growing

This chapter is by way of brief interlude: a pointer to some of the complexities of action and policy which marked the progress of ICSU during the five decades which are dealt with in the following chapters.

There is no simple index to the growth of science, although we are all aware that every year there seem to be more and more books and journals, and more and more people earning their livings in activities we think of as scientific. Derek Price's well-known study *Little Science, Big Science*[1] was the most widely read of many efforts to measure scientific activity, and its influence remains to remind us all to keep an eye on numbers.

We should need to know about numbers of university chairs in various subjects, numbers of students of undergraduate, graduate, and post-doctoral status, membership of learned or professional bodies, numbers and value of grants made by government and charitable granting bodies in those countries which have them, numbers of scientifically graded employees of science-based industries.[2]

In the immediate post-war years the active participants in ICSU (who were not really very numerous) had liked to think of themselves as having roles on a world scene. This was an expression more of hope than of real achievement. By 1963 the work that stemmed from ICSU had become truly global, territorially, financially, in influence, and in scientific activity, providing substantial justification for this claim to world status. That early justification lies largely in the programme called the International Geophysical Year, in the setting up of the International Biological Programme, and in other programmes for which these served as models.

The generalities about freedom for science in the Hale pronouncement of 1934 and in the 1958 General Assembly resolution would look thin if they had expressed some conventional view, but they were as much the embodiment of hope and ambition as of moral conviction, a hope and ambition that ICSU would grow freely in an increasingly free world. And grow it did. The national membership increased from 40 in 1931 to 82 in 1994 as a result of constant acquisition especially in the developing world. Also, after the end of the Cold War, the number of national members rose steeply when the old academies of the countries that until then had been represented by the USSR National Academy began to admitted in their own right.

The growth of Union membership was less spectacular but gradually filled in gaps which had existed for a long time, and there were being added entities representing totally new departures in science.

New Unions

There were 7 Unions in the ICSU family in 1945. In 1993 there were 20. Why were new ones needed and how were they formed?

We get a good picture in the case of X-ray crystallography. IUCr, the creation of which was first discussed in 1946, was brought into existence because of the development, not of crystallography, which had its origins in the 17th century, but of X-ray crystallography which originated in about 1913. By 1946 there were enough X-ray-crystallographers in the world and in touch with each other for them to want to combine. Moreover, although publication was important, a mere learned society would not quite meet their needs. The reason for this was that their subject was already proving useful in throwing light on problems in other fields of science pure and applied. A Union, with its ICSU-guided links with other Unions, was a better form.[3]

Those who wished to form a new Union did not always have it their own way. The chemists (IUPAC) thought biochemistry should be viewed as a branch of chemistry and should therefore be organised as a section of IUPAC.[4] In the end the biochemists, who had developed their own community sense, had their way and IUB was admitted to ICSU in 1955.

A Union somewhat removed from the main field of experimental science marked a different point. As we saw in chapter 6, the year 1946 saw the admission of a Union which had existed in other forms for some time,[5] the International Union of the History of Science. Formation of a Union did not mean automatic membership of ICSU, as one sees from the International Union of the Philosophy of Science formed in 1949 but not immediately admitted as a member of ICSU. It was later admitted when it federated with IUHS in 1956, functioning thereafter as one of two Divisions of IUHPS. This is quite typical of the ways various Unions have found of arranging internal organisation to suit the needs of parts of a wide-ranging area of science which function happily and appropriately together. On paper, the historians and philosophers of science used, as it were, dialects of one intellectual language, but their vitality came from speaking as individuals, face to face, the common language of everyday.

IUBS suffered considerably from the formation and subsequent admission of new Unions in specialised fields of biology, such as the International Union of Immunological Societies (IUIS) (1976) and the International Union of Microbiological Societies (IUMS) (1982). Some saw this as a weakening of IUBS, others feared that the increasing number of biological Unions would upset the balance of power. Yet others, probably rightly so, recognised that the promotion of a well-defined part of biology such as microbiology could best be assured by a specialised independent Union rather than as a sub-division of a large 'Mother Union' to which linkage would tend to be formal only. On the other hand a Union like IUGG (1919) has so far withstood any centrifugal forces that may have threatened it. To date it comprises no less than seven different associations.

Another area of centripetal development is seen in IUTAM (International Union of Theoretical and Applied Mechanics), created in 1946. Its objective was to 'create a link between persons and national or international organizations engaged in scientific work (theoretical or experimental) in mechanics or in related sciences'. What this means is seen in the field of interest of some of the organisations which became affiliated to IUTAM in later years.

1970 Mechanical Sciences (CISM)
1972 Heat and Mass Transfer (CHMT)
1974 Rheology (ICR)
1978 Vehicle System Dynamics (IAVSD)
1978 Interaction of Mechanics and Mathematics (IMM)
1978 Fracture (ICF)
1982 Mechanical Behaviour of Materials (ICM)
1982 (Asian Congresses of) Fluid Mechanics
1984 Computational Mechanics

These are all subjects each of which would have been the occupation of a rather small number of individuals in the first decades of the century. By the post-war period they had each been studied by large numbers of specialists, often with the stimulus of industrial or manufacturing problems.

Nutrition was also being widely discussed long before the war ended. The stresses of war made people think in new ways about familiar things including nutrition. This was certainly an old concern, the relation between food and wellbeing having been brought into the laboratory at least as early as the 18th century, with increasingly complex studies in the chemistry of nutrients and their use growing through the 19th century. (One finds in the roll of honour of these studies names like Lavoisier and Liebig.) Much of this was necessarily on the principal structural and energy-providing factors in diet, but with the enlargement of the known range of necessities (vitamins, for example) nutrition became a science of itself. It had the distinction also of appearing to be interesting not only as pure science in itself, but also socially and economically necessary. Between the wars there had been several barely successful attempts to bring the understanding of the basis of nutrition home to ordinary people as well as to economists and politician. Boyd-Orr[6] was a leading figure in this, but he was not alone in finding it hard to persuade governments of its real significance. During the Second World War pressure on foodstuffs supply at last made the study of nutrition of great economic significance. Afterwards the time was ripe for the formation of an International Union of Nutrition. Discussion in 1946 led to the formation of a body which was eventually elected to membership of ICSU in 1968, as the International Union of Nutritional Sciences.

There could not be a better example of the expansion of science than the formation of IUB (International Union of Biochemistry) to which we have referred. It had a somewhat difficult birth, since the chemists in IUPAC, one of the oldest Unions, had thought that biochemistry should remain within the IUPAC fold. However, a separate Union was formed in 1955. Before long the work of the X-ray crystallographers was having an effect on the study of some very fundamental biochemical matters, and it became clear that a new science of molecular biology had to be embraced by the biochemists. IUB became IUBMB (International Union of Biochemistry and Molecular Biology) in 1991.

The geologists already had a long tradition and experience of international activity, having met in International Geological Congresses since 1875. Geological studies had changed bit by bit, and it was recognised that many geoscientific problems were of world-wide significance and required world-wide study, best achieved by a new style of organisation. IUGS (International Union of Geological Sciences founded in 1961, in which year it adhered to ICSU) recognised also that many of its interests had economic, industrial, and environmental relevance, and the very large number of committees and commissions it eventually set up reflect this diversity.

Another group with a long tradition of congresses was that of the physiologists, who had been holding international congresses since 1889. The subject was less clearly differentiated in its early days; biochemists and pharmacologists played some considerable part for a time, but they gradually formed their own communities. A Permanent Committee was established in 1929 to organise the physiological congresses. This became an International Union of Physiological Sciences (IUPS), adhering to ICSU in 1955.

Until the period of the Second World War, pharmacology had been practised as a supportive technique for some branches of medical research, confined mostly to the measurement of the effect of drugs or chemicals on biological systems. With the expansion of the study of synthetic drugs and the elaboration of physiology, pharmacology became a study in itself, with an expanding body of practice and theory and a corresponding growth of societies devoted to its study. The International Union of Physiological Sciences had recognised this growth by the creation of a Special Committee for Pharmacology, but in 1965 this section seceded and created its own International Union of Pharmacology (IUPHAR). It organised its own congresses, established its own image, and was accepted as a member of ICSU in 1972. It was also at same time accepted as a non-governmental member of WHO, a fact which illustrates the complexity of ICSU's links with the network of intergovernmental scientific activity.

Such a change is not as simple as it looks. IUPS, as its President observed in announcing the secession to the IUPS General Assembly in Tokyo in 1965, had to make a number of revisions of its Statutes and to consider how it was to collaborate in future with a body which had hitherto been one of its own components.

Another factor, not explicit in the constitution-making process and by no means obvious to the outside observer, was that IUPHAR's subject-matter was one of those which were likely to bring ICSU into closer touch with industry. It was within the pharmaceutical and chemical industries that much of the increasing impact of pharmacology on the public was being generated. The reasons for the emergence of new disciplines were not always purely scientific: the growth of industry was having an effect on ICSU, even though it had not yet found any general way of bringing industry into partnership. (It was still seeking one in 1990 at the Visegrad Conference.)

Another new Union had 'pure and applied' in its title. An International Organisation for Pure and Applied Biophysics was established in 1961, becoming a Union (IUPAB) and adhering to ICSU in 1966. One needs only a single example to highlight the broad scope of its interests: that one of its affiliated commissions is a group of the Institute of Electrical and Electronic Engineers.[7]

An example of an old medical practice taking on a more detailed scientific aspect in the 20th century is immunology. The International Union of Immunological Societies was established in 1969. The simple immunisations that had been discovered in the 18th century had become, by the mid-20th, a very widespread permanent feature of health care. Moreover, the understanding of the immunity systems in relation to other things than protection against specific disease had become a major academic study and a necessary feature of much major surgery. As mentioned above, IUIS became a member of ICSU in 1976. That their subject was not one isolated by academic detachment but has a day-to-day force is shown by the magnificent WHO achievement of the eradication of small-pox by 1979.

No two Unions developed in the same way. The International Union of Microbiological Societies began in 1927 as an International Society of Microbiology, was then affiliated to IUBS, (under the name of the International Association of Microbiological Societies, IAMS) as a Division in 1967, became independent in 1980, and was admitted to ICSU as a Union member (IUMS) in 1982.[8]

In 1982 also the International Union of Psychological Science (IUPsyS) became a member of ICSU, nearly a hundred years after the first International Congress of Psychology had been held in Paris (1889).

For another 11 years no new Unions were admitted to ICSU, although some bodies gradually increased the warmth of their relationship to the point at which Union membership seemed appropriate and was sought. Three were admitted in 1993. IUAES (International Union of Anthropological and Ethnological Sciences), ISSS (International Society of Soil Science) and IBRO (International Brain Research Organisation).

Although the size and character and also the resources of the 23 Unions were very different, as each had its own history, the scientific scope of ICSU widened over the years to the point where almost no significant branches of science were lacking. It is furthermore remarkable that 4 out of the 23 are considered to belong to the 'social sciences', that is IUHPS, IGU, IUAES, and IUPsyS. Direct representation of the medical, engineering, and agricultural sciences was weak. With regard to the latter a beginning was made by the admission of the International Society of Soil Science in 1993.

New bodies

Up to 1952 ICSU was growing through new national members and the admission of further Union members. Academies and Unions can exist very well by themselves without ICSU. What then caused them to join together? ICSU offered them the opportunity for co-operation in fields that engaged the interest of more than one Union and in addressing world problems that required effort from various branches of science (e.g. fresh water, oceans, global environment). Furthermore ICSU could and did address concerns common to all sciences and scientists, such as free circulation, teaching of science, and publishing. Finally ICSU was seen, as the membership grew, more and more as the 'spokesman' of science. Thus from 1952 onwards its structure and relationships began to develop a greater complexity through the creation of new bodies for purposes which were only now beginning to emerge.

We could put these reasons under some rough and ready headings: communication, co-operation, synergism, innovation, intervention, education, but we should find more than one of them in operation with every example. Undoubtedly the most important initiative created was the IGY, implemented in 1958–9. It was to influence the work of ICSU in many ways during the second half of ICSU's existence and merits a chapter on its own (chapter 12)

COSPAR, set up by the Eighth General Assembly of ICSU in Washington DC in 1958, was a continuation of an entity which had come into existence during the International Geophysical Year. Not only was IGY the largest international scientific enterprise yet carried out, it had produced some entirely new situations and challenges. Perhaps the most striking of these was the introduction of the rocket as a routine vehicle for scientific apparatus, carrying investigation far beyond the limits of the atmosphere.

Another continuation of the work of IGY was SCAR (Scientific Committee on Antarctic Research) formed at the Hague in 1958. Apart from the value of work agreed between scientists in their own professional capacities, SCAR is distinctive in its responsibility to provide scientific advice to the Antarctic Treaty System. Twelve nations were involved in the major work and found the results so valuable that they agreed to continue to promote scientific work in the area, unhampered by political considerations. A treaty was signed in Washington on 1 December 1959, coming into force in 1961. A protocol signed in 1991 records agreement to protect the environment and to prohibit mineral exploitation for at least 50 years. The consequences of IGY therefore go far beyond those of any such scientific endeavour before its time. (We see this later in the formation of the International Biological Programme, IBP.)

We shall deal with the origin of FAGS, the Federation of Astronomical and Geophysical Data Analysis Services when we deal with IGY, but here it is worth noting the range of permanent services that some of its leading members (IAU, IUGG, and URSI) undertook: earth rotation, geomagnetic indices, gravimetry, solar activity, mean sea level, earth tides, Ursigrams, glacier monitoring, stellar data, sunspot data.

The setting up by the Executive Board of ICSU of SCOR (Scientific Committee on Oceanic Research) (1957) is a reminder that the oceans are no respecters of national boundaries or of national or commercial interests. Their study presents a variety of problems nearly all of which have an international face. Quite apart from the technicalities of experimental methods, which benefit from wide discussion, there is strong reason to design co-operative experiments and surveys. SCOR has gradually increased the range and reach of its activities so that it progressively extended the scope and planned duration of its programmes. The constitution of SCOR necessarily involves nations through representative national institutions or laboratories, and the promotion of the scientific standing of laboratories so that they may have appropriate authority. With such a concern for national activity, SCOR is able to present the views of marine scientists to national governments and to intergovernmental organisations.

There are other broad views. Water is everyone's concern: a fact reflected in the formation in 1964 of COWAR, a Committee on Water Research of ICSU itself, and its subsequent transformation in 1976 into a Joint Committee of ICSU and the International Union of Technical Associations (UATI). Through UITA, ICSU tried to involve itself in the work of practical engineers, active very often in those fields of environmental work which ICSU approaches also in other ways. In 1993 COWAR was transformed into a Scientific Committee, SCOWAR, with an improved mandate. Since the formal link with UITA had not proved as effective as initially hoped, it was abandoned and replaced by informal co-operation on a case to case basis.

One new creation by ICSU was appropriately carried out at that first General Assembly outside the western world, in Bombay in 1966. This was the setting up of a Committee on Science and Technology in Developing Countries. By 1990 it became clear that it should do more than originally envisaged, so COSTED was reorganised to function also as an advisory body on all the activities in developing countries. One of the progressive features of the new COSTED structure was the creation of regional Secretariats in Asia, Latin America and the Caribbean, and in Western and Central Africa.

From its inception, COSTED slowly built up local centres of education and training, using scientists from both inside and outside the regions.[9] One problem is the rate at which scientific techniques changed, so that the young scientists of a region had to be taught both what was accepted practice and what was being introduced in the world at large. Another programme, the International Biosciences Networks (IBN)(1979) was carried out by ICSU, jointly with UNESCO, in Latin America, in Asia, in the Arab Region, and Africa, in the latter case with support from UNEP. Since IBN and COSTED had a great deal in common, a merger was effected in 1993 leading to the COSTED/IBN.

In contrast to the broad generalisation of the COSTED concept, the same Bombay General Assembly set up SCOSTEP (Special, now Scientific, Committee on Solar-Terrestrial Physics). This always had an inter-Union character and aimed to promote interdisciplinary programmes involving at least two of the co-operating Unions: IAU, IUGG, IUPAP, URSI, together with COSPAR.

The purpose of this body (SCOSTEP) was to promote the kind of interdisciplinary programmes to which the variety of interests in the sun very naturally led. There was no continuing programme. Each specific programme was of finite duration, and it was planned that the resulting data would be dealt with by the World Data Centres. The setting up of SCOSTEP was a further example of ICSU's role in creating machinery for bringing different areas of science into fruitful contact with others. Early in its

history it might have looked as if the expansion of ICSU would come about by the creation of new Unions. In fact it happened as much, if not more, by the creation of new
bodies for purposes which could not be contained within the purposes of one Union
alone.

Another action of the 1966 Bombay General Assembly (it deserves to be referred to
as an achievement) was the setting up of CODATA (Committee on Data for Science
and Technology). It had its origin in the review of the work of the World Data Centres
that had been set up during IGY. It eventually became an important body in its own
right.

These World Data Centres developed more or less individually after the success of
IGY had shown the value of co-ordinating data collected in many countries. Soon
however it became apparent that much more could be done than the un-coordinated
collection of data. ICSU set up a Panel on World Data Centres (Geophysical, Solar,
and Environmental) in 1968 to advise on the management of the many centres (now 40
in number). The gathering together of data from many sources is a valuable means of
monitoring changes in the geosphere and biosphere.[10]

When our descendants look at the 20th century, they may well distinguish it from
other centuries as having two characteristics: the attempt to establish means of ensuring world peace, and the reassessment of *humankind* as *humankind in a vulnerable
environment*. One used to think of the human creature as being vulnerable in a stable,
if hostile, environment. In the last three decades we have begun to realise that it may
be the environment which is vulnerable, in the face of a hostile humanity. In the sixties,
not only on a national scale, but also internationally, several organisations were set up
to address the growing concern about the effects of human actions on the global
environment on which humankind depends, on a more extended scale in space and
time than hitherto encountered. UNESCO started its Man and the Biosphere
Programme (MAB) and the UN convened in 1972 the UN Conference on the Human
Environment in Stockholm, which led to, among other things, the setting up of UNEP.
Instinctively ICSU found its correct role: scientific analysis of the problems, pulling
together available pertinent knowledge from different disciplines, and development of
research agenda and of unbiased advice to governments. In 1969 the Executive Board
of ICSU created, in its meeting in Erevan, the Special Committee on Problems of the
Environment (SCOPE) with the above tasks in its mandate. Like the IGY it had great
influence on future ICSU activities and showed *par excellence* the synergistic effect of
bringing to bear a number of disciplines on the analysis and solution of major complex
problems.

SCOPE was but a prelude to other major activities: the WCRP (jointly with WMO
and later also with IOC), the IGBP and the Global Observing Systems (GCOS, GOOS,
and GTOS) (1992–3), together with various UN bodies.

Concern for the good of humankind in the face of a new danger was also a leading
element in the existence of COGENE (Scientific Committee on Genetic
Experimentation), which was formed at the Sixteenth General Assembly of ICSU in
1976, to serve as a source of advice concerning recombinant DNA activities.

The pure science of this phenomenon was already very great in extent, but the variety of problems in benefits, environmental health safeguards, and containment were also clearly of such magnitude that some international scrutiny was needed. DNA was the focus of many kinds of inquiry, at first in pure science only, but rapidly moving on into applications full of hope and of fear.

In its formal session, the 1974 General Assembly heard a little of the concern being felt at the emergence of a new field of science following early work on the molecular basis of the mechanism of heredity. There was also serious informal discussion in the corridors, in which J C Kendrew took a lead. By the time of the 1976 General Assembly, Kendrew was Secretary-General and was able to bring this topic to prominence in the Agenda. W J Whelan gave a lucid lecture explaining how it was now possible to modify the heredity of some species so as to produce what were virtually new creations. The bright prospect for scientific discovery was matched by the dark prospect of danger through the inimical new organisms. It was agreed that ICSU should give this topic high priority, and set up a committee which could call on the best brains. Thus COGENE was set up to provide a forum through which the many interested bodies, national, international, and (increasingly important) regional may communicate. It worked effectively on its original basis, and by 1993 a whole new area of applications of the new discoveries on an industrial scale had come into existence.

These developments played a part in the wider front of biotechnology, progress in which led to the setting up in 1986 of COBIOTECH (Scientific Committee for Biotechnology). This was to promote biotechnology for the benefit of humankind world-wide, through research, education, and communication. It aimed also to reflect ICSU's concern for developing countries. COBIOTECH had a technical outlook, seeking to make the common experiences in biotechnology widely available for the material benefit of all. Here again we see the gradual acceptance by ICSU of responsibilities for applied science of a range which the ICSU founders would have found unfamiliar, to say the least. By 1993 it was proposed that COGENE and COBIOTECH should combine, but this had not come about by the end of 1994.

In these new bodies for special purposes we can see one important feature of the growing influence of ICSU: its reach into applications of science to useful purposes. No human activity could be more manifestly utilitarian than getting good out of earth and water. The phrase 'The Good Earth' has a poetic resonance that expresses our absolute dependence on the endless cycle of growth and decay. Committed to this concept was the decision of the Seventeenth General Assembly (1978) to create an Inter-Union Commission on the Application of Science to Agriculture, Forestry, and Aquaculture (CASAFA). It became a Scientific Committee at the Twenty-Third General Assembly in 1990. It is a pity so clearly desirable a body should have such a ponderous title, but its aims were fundamental: to promote co-operation between institutions and individuals in developing and in industrialised countries 'towards the resolution of problems of food, feed and fibre production and towards the reduction of post-production losses'. This business-like phraseology is a long way from the academic abstraction of the dedication to unsullied pure science of the forerunners of ICSU earlier in the century, but its practical achievements seem not to have been in keeping with its intentions. Its activities went on until 1993, when it was argued in the General Assembly at Santiago that ICSU's concern for agriculture and ICSU's potential contribution needed to be analysed in the light of prevailing conditions. CASAFA was suspended until ICSU could reformulate its role in agricultural science and industry.

It is easy to recite examples of the comfortable material benefits of science in easing deficiencies or of relieving pain, but some of the threats to life can bring extreme deprivation propelled by terror. Let us end this survey with the committee ICSU has formed which faces 'the worst that can happen'.

The hazards of environmental change are one thing: the perils of natural disasters are quite another, but, in spite of their apparently random nature, they can be studied scientifically. The phenomena involved in volcanic eruptions, in earthquakes, or in tropical storms are of challenging interest to many Unions, so we find that a recent ICSU creation (1990) is one related to a United Nations project: IDNDR, the International Decade for Natural Disaster Reduction. The parallel ICSU Special Committee SC-IDNDR has set up projects for all three of the kinds of natural disaster mentioned, and has an educational programme as well.

It might seem as if this committee would be hampered by the very randomness of its *raison-d'être*, but this is not the case. Its plan was to concentrate on those natural disasters which are of an acute (i.e. relatively short-term) character, as opposed to those which develop over long periods. Its objective was to reduce the human impact of disasters through knowledge of their causes.

Unlike some other committees which have no *terminus ad quem*, SC-IDNDR aimed to cover the UN decade. It had a very mixed composition, with members from IGU, IUGG, IUGS, and IUTAM, as well as expert individuals. It also invited representatives of WFEO (World Federation of Engineering Organizations) and UATI (International Union of Technical Associations and Organizations).

The 'Spearhead Projects' are studies of Tropical Storms, Volcanic Eruptions, and Seismic Hazards. There has been success in methods of study of storms and in prediction of eruptions, but less progress in seismic prediction.

There are two Second Wave projects, one old and familiar in form (Famine), one a creation of the 20th century (Megacities). Famine can occur for many reasons, known since ancient times. The modern megacity is vulnerable for many reasons, its presentation of a large target for any physical hazard, the complexity of its essential services, and its sheer mass, weighing down and imperilling the very earth structure on which it stands and which may fail to support it.

Common concerns

The effective teaching of science is everyone's concern, and many Unions put some considerable effort into this duty. An Inter-Union Commission on the Teaching of Science (CTS) was set up in 1961, and modified to a Scientific Committee by the

Twelfth General Assembly in Paris in 1968. It was concerned with the teaching of science at all levels, and therefore found it useful to have contact not only with unions, but also with other bodies which have the raising of effective standards as their aim: COSTED and ICASE (International Council for Science Education).

In 1993 CTS was replaced by the Committee on Capacity Building in Science (CCBS) which had a twofold mandate: promoting the building up of scientific capacity in all parts of the world, and enhancing, among the public and decision-makers, the understanding and appreciation of the proper role of science.

Other important ICSU bodies addressing common concerns are, of course, the Standing Committee on Freedom in the Conduct of Science (SCFCS) described in chapter 8 and the ICSU Press Publishing Service (1983, 1989) which advises ICSU family members in publishing matters.

This description of the evolution of these many bodies has necessarily treated them as rather abstract entities. The people who created them were composed of as much human strength and frailty as any other people. There are probably as many tales to be told of ambition, position-seeking, obstinacy, and obscurantism as there would be in the history of any other considerable organisation. The success of ICSU suggests that they would be outweighed by many other tales of ambition for the good of ICSU or Unions, and obstinate determination to overcome difficulties in the path of shared success. The one quality that seems to have been transferred from the pursuit of science to the development of ICSU bodies is inventiveness.

ICSU has used any machinery of organisation it found convenient for any particular purpose. If so condensed a recital of formal and informal bodies seems confusing, the reason is that life is like that, and ICSU is concerned with life as lived, which is why ICSU has changed. When it began, its outlook was limited by its having to rely on a few people in a few organisations in a few countries. Outside events and its own growth have combined to make its end-of-the-century outlook global in authority and global in scientific reach.

Chapter 11

ICSU at mid century

Authority building

Too many things happen at once to any organisation of the size of ICSU for one to be able to recite a simple chronology. (Is this not true of any family history?). But we can pick out particular features of ICSU's work, and see where their origins lay. We cannot confine each topic to any period, even to a particular decade, because none of the elements in the history of ICSU had the same kind of lifespan. Obviously ICSU in 1994 is not the same as ICSU in 1945. It had changed in number of members, in kind of members, and in many of its objectives. The period 1945 to about 1975 did, however, see a number of major changes we can identify as characterising one generation. The changes in the period 1975 to 1995 have not yet taken on as clear a form, but we can see them leading up to a period of internal re-examination and of adjustment to great changes in the political world.

Some bald statistics to begin with: in 1950 the United Nations had 59 members, UNESCO 57, and ICSU 43 national members. During 1960 the United Nations grew from 82 to 99 members (much of the addition in that year being African states). UNESCO also had 99 members (but some of these were not members of UN). ICSU had 48.

Organisations are meaningless without people: individuals and kinds of people. Much of the inner strength and public image of ICSU, as in any other body great or small, lie in the kind of people who can, if need be, function effectively as its officers. From its foundation in 1931 until 1963 all the ICSU officers came from either Western Europe or the United States. In 1963 D Blaskovic of Czechoslovakia became Secretary-General, the first step to the broadening of the base of authority. In 1966 he was succeeded by the Indian K Chandrasekharan. The first President from the 'developing world' was D A Bekoe, of Ghana, elected 1980.

The President marks change and is its chief maker. The Secretary-General is sometimes its chief agent, the extent to which he can initiate change depending on the character of the President, and on their good relations.

The period from 1945 to 1958 was the transitional period, referred to in chapter 5, in which the Secretaryship was maintained by devoted senior officers of the Royal Society who used their influence to develop a more flexible outlook. Since the formation of ICSU, they had all operated from private College rooms. By the end of this Royal Society period (under the Secretaryship of F J M Stratton[1] and then of A V Hill),[2] operation from a permanent office with a paid staff secretary became possible, largely because of the new link with UNESCO. The accommodation was at first provided by UNESCO in Paris, and a new style of continuity was achieved. This was found unsatisfactory, and the Royal Society negotiated accommodation in rooms lent by the British Museum (Natural History) and then in rooms leased from the University of London until 1958. In 1958 the office moved to the Palais Noordeinde in The Hague.

Up to and including the time when the Swiss von Muralt was President, ICSU had to rely for the presentation of accounts on the Secretary. In 1952 when B Lindblad (Sweden) was elected President there was elected also the first of a line of Treasurers who were able to concentrate on the money matters which became more and more substantial and more and more complex. W A Noyes[3] was the first.

Outstanding in service to ICSU was Georges Laclavère. He was one of the most experienced servants of ICSU. He had long experience in IUGG and in other bodies, national and international).[4] As Treasurer he set a pattern of being re-elected to serve three terms (1961–8), which also happened to N B Cacciapuoti (1968–72), T F Malone (1978–82), and K Thurau (1984–8)).

The effect of the imagination of a President is to be seen notably in the work of men like Lloyd Berkner in the initiation of IGY, or Harold Thompson in the creation of COSTED, but each President made his own contribution. One died in office (E F R Steacie) before he could fulfil the promise which led to his election.

On the other hand, important ICSU initiatives could originate equally well from any member of the ICSU family, from a Union, a national member, a committee or any Executive Board member. There were – and are – almost always some new ideas floating in the air, some of which gather support and eventually come to fruition.

Each of the earlier Unions was founded by adherents of a particular discipline who saw a need to work together for more than could be provided by any of the national associations or societies. Many of these Unions had already grown up, for meeting, talking, and publishing. Not only had *ad hoc* international meetings existed for a very long time; some international organisations had a long tradition of providing for an organised succession of congresses. The Union type of constitution, however, went further, put more certain order into the congress system and extended the facilitator function. Their principle was that a succession of open meetings could be carried out by a body which was also able to organise other kinds of work, most strikingly on long-term problems, on co-operative ventures, and in common services. Continuity was added to succession. To use a fanciful image, a mere succession of congresses could be seen as the islands in an archipelago. Under Union control they were more like peaks in a mountain range. The Unions also addressed the question of bringing order into the scientific language of a discipline in which its activities and findings could be recorded or communicated. This was another powerful instrument for continuity and uniformity.

ICSU had no monopoly of the word 'Union' (as demonstrated by the autonomous

International Union for the Conservation of Nature, later renamed World Conservation Union) but its Unions were achieving a certain distinction of style by the time the Second World War broke out, and all could claim certain achievements. The common virtue was that they had improved communication between participants in the scientific skills they represented.

We looked earlier at some of the advances in science and technology that made the late thirties so interesting a period. The first years that followed the 1945 relief of the tensions of actual physical danger had inevitably to be concerned with consolidation, with the picking up of threads, the revival of memories of how to do what had been ordinary things, the hardship of ruined economies in most combatant countries. The housewife in her kitchen felt this hardship more than anyone. But manufacture and trade gradually returned, not to any supposed 'normality' but to new technical conditions. It has often been pointed out that the state of the motor car and its industry is a good index of economic and technical progress in the consumer market: the production of cars for a civilian market opened up again with models which showed no new features for some time. A huge aircraft industry grew up. The pharmaceutical industry expanded. Housewife and entrepreneur had to replay old programmes for some time before finding that many of the old ways had gone. However, pure science had fewer restraints; although it had suffered from wartime distortion of objectives, it benefited from wartime discoveries and techniques, and could change faster than civilian pursuits.

In the world-wide state of uncertainty of the late 1930s, ICSU had maintained sufficient stability to be ready, had there been no war, to embark not only on new activities, but also on increased membership, scientific and national. After five years delay a new start could be made, but there was then in very many areas (of thought, of action, of government) a radical realignment brought about by the war. ICSU expanded its scope and authority in some ways. A few were in evidence before the war, many others came to the fore afterwards. Some expansion was due to an increase in scientific activity; some was due to changes in world outlook, even more radical in its political and social features than could have been envisaged.

What is true of people is sometimes true of organisations. If any body or party or person is to act effectively, it must look authoritative: ICSU had shown a degree of authority from its earliest beginnings; some of its constituents had brought their own authority into the creation of IRC and ICSU itself. An outstanding example is that of the work of IAU and of URSI[5] in those fields of observation and communication, of mapping, and of the regulation of time on which world order so much depends. Since the late Middle Ages, people learned to live by the mechanical clock, content with local time until the coming of the telegraph and the railway timetable. IAU and URSI have helped us live by a world clock. IUPAC contributed to the regulation of the language of chemistry and of its critical data. It also played a part in the harmonising of pure and industrial chemistry.[6] This was of the greatest importance in chemical industry, which above all others converts materials useless in themselves into useful substances, thereby increasing overall human material wealth. If we are to suggest a date when ICSU took on greatly increased authority let it be 1958, that of the opening of the

International Geophysical Year (1958–9). Its effects were great and many, but it was not the only stimulus to progress.

From 1958 onwards we find international activity increasing both in what was centuries old and in what was new in the present century. Less than 20 years after the events of 1945 there were many new international enterprises in being of kinds which went beyond territorial and national limitations. An ancient field of study with a long tradition is exemplified in modern form in two International Oceanographic Congresses held in 1959 (New York) and 1960 (Japan). The change of venue makes its own point about the gradual calming of enmities. In 1959 an International Cartographic Association was founded (it joined ICSU in 1990 as a Scientific Associate). Its aim was to raise standards of technique and of the quality of cartographic practice. This spread of international activity helped create the atmosphere in which ICSU could lead and demonstrate its authority.

The new is exemplified about the same time (1960) by a foundation which reflects the recognition that scientific activity has often to be regulated by agreement. IUCAF (the Inter-Union Commission on Frequency Allocation for Astronomy and Space Science) was set up by URSI, IAU, and COSPAR. It made it possible to advise national and other bodies responsible for frequency allocation within their several jurisdictions on the best way to accommodate the needs of scientific radio in harmony with the needs of commerce, industry and official communication. (The problems remain, different in detail but undiminished in gravity, thirty years later.)

In 1961 a late-comer to ICSU was the International Union of Geological Sciences. Geology had been actively organised with well-managed, well-attended congresses every four years since 1875. There had been many comments in ICSU General Assemblies about the absence of geology from the roll of ICSU members, but the gap was now filled. It was a natural thing to happen, because the links between the sciences were becoming more numerous and firmer, with many studies being carried out in which the earth itself was the material and the setting. Co-operation between geologists and colleagues was made all the easier by this enlargement of the ICSU family.

The Twenty-First International Geological Congress held in Prague in August 1968 was disrupted by the invasion of Czechoslovakia by the Warsaw Pact forces, but resolutely carried on its work as much as possible and eventually managed to achieve an excellent post-Congress publication programme.

None of this happened overnight. These changes of state and affiliation all took time, sometimes influenced by changes in the membership of committees or their leadership. Pride and prejudice played their part. For example, 1961 also saw the formation of an organisation which reflected that clarification of fields of study which is constantly taking place at the same time as boundaries between disciplines become blurred. The International Organisation for Pure and Applied Biophysics was founded at a meeting in Stockholm in 1961. It later sought membership of ICSU (with the title of Union, IUPAB), which was accorded by the Bombay General Assembly in 1966. Some members of IUB speaking at the ICSU General Assembly of 1966 in Bombay were not entirely happy about the recognition of such a body as distinct in function. A peace-making request from the officers of ICSU that IUPAB should establish close links with IUB was met by the reassurance that such links had been established already for some years. The voting was 53 for, 9 against, with 10 abstentions.

This was the same General Assembly at which an application for admission from the International Union of Nutritional Sciences was turned down after a long and complicated debate about the nature and significance of studies in nutrition. It was argued that the nutritional sciences were adequately represented by the disciplines of existing member Unions. Two years later the General Assembly of 1968 in Paris thought again and admitted IUNS, the voting being 50 for, 19 against, with 3 abstentions. This change of heart after so little time may look absurd, but one of the things that lay behind the original rejection was the belief that the authority of ICSU should not be diluted by having its affairs managed by too unwieldy a voting body. The internal politics had to be sorted out and separated from the scientific logic. Misgivings had been talked out by 1968.

It is not necessary to recite the conditions and voting that led to each and every admission. This one example should explain that the growth of ICSU membership was not automatic or unregulated. Growth of membership was allowed selectively only so far as it brought with it a growth in authority. Authority arises also from things knowingly not done. During its entire lifetime ICSU has been careful to distinguish itself from many other non-governmental organisations (especially in the field of environment) by never acting as a pressure or advocacy group. It has always restricted itself to presenting scientific facts in an objective and dispassionate manner, thereby maintaining the high level of credibility it has acquired over the years. Authority does not provide power, but it does lead to influence. At the mid-century ICSU was already the most influential international body of its kind.

Regional problems

Another kind of achievement of the post-war years was the recognition of, and adjustment to, regional problems. Before the war there had been a certain amount of work generated by Unions in small areas of the world, often because of the particular interest of local biological or geophysical characteristics. However, this work had been mainly done by visitors, and expatriates. Bit by bit, as new national members adhered to ICSU, it became easier to adopt an attitude of co-operation in place of one of charitable patronage.

Eventually something wider in scope emerged, as more became known about the needs and aspirations of countries in course of development, which by no means followed the historical pattern of the advanced countries. Sometimes the introduction of 'Western' technology had been of great benefit to a country, as in the introduction of railways to India. However, the fundamental agricultural, nutritional, and social patterns of many countries were so different from those of what was now often called the 'North' that development had to be considered in new terms.

This is illustrated by some instances of failure of good intentions to promote astronomical work in some Third World countries. Excellent telescope equipment remained in its packing cases until out of date. It can be said that Africa needs agricultural knowledge more than it needs astronomy, but this is too simple an explanation. The development of scientific activity needs an infrastructure which cannot be delivered in packing-cases, but has to be added to by the training of new recruits to a growing scientific community, supported by, or at least tolerated by, the administration.

ICSU is not alone in facing this problem. There is an example in some work of the International Council of Museums. Ghana was anxious to develop its universities, but its academic leaders were also aware of the need to develop a scientifically enlightened body of popular support and a technical work-force. A proposal to form a Science Museum in Accra made good progress in 1963 until it was brought to a prolonged halt by political events,[7] but the concept was correct: skilled mechanics are needed as much as engineers, but you cannot create by legislation a social pattern in which there are all the levels of training and skill that characterise a mature industrial society. Putting it that way exposes the need to decide what kind of pattern of skills and training is in fact appropriate to any particular country or region. In the 1980's the phrase 'Capacity building' came into use but at the mid-century it was not yet part of the vocabulary of goodwill.

Wealth plays its part, but not necessarily a leading part, in development of scientific self-consciousness. Nigeria adhered to ICSU in 1963, a little before the time when exploitation of oil off its coast precipitated a succession of economic and political upheavals which continue to the present day. However the scientific community has maintained its viability and a Nigerian National Academy of Science was founded in 1977.

By the 1960's one could recognise a general division of the world into three economic styles, each distinguished by a characteristic management in each country of its economic affairs: the capitalist group (no longer merely a geographical West but linked also with some countries around the Pacific), the communist countries (economies managed by an ideological political system), and the 'Third World' (countries characterised by a GNP lower than that of members of either of the other two groups). The capitalist and communist countries engaged for a long time in a military confrontation (a 'Cold War')[8] which the scientific communities of each managed, very often and effectively, to ignore when it came to questions of scientific communication.

The term 'Third World' came into use soon after the Second World War. The countries to which it was applied (mainly in Africa and South America) were very different from each other, but the problems they posed the scientific community were similar enough for it to be useful to plan some degree of common approach to them, with the general view of promoting development.

Other kinds of change were taking place, for quite different reasons from those which applied to the Third World. In 1961 both Koreas (Republic of Korea, Korean People's Republic) adhered to ICSU, eight years after the military end of the wasteful war of 1950–3. Whatever the military and the politicians might have to say to each other, at least the scientists of both countries sought some common ground, and a forum.[9] The Korean conflict had the effect (among many others) of reminding the 'Western' world that their 'East' was not just Japan and China.

Geographically the Third World countries lay for the most part in the tropics, which gave a certain unity to thought about it. There was another kind of geographic unity becoming apparent: the formation by some of the countries of the Pacific basin of the Pacific Science Association in 1920. After 1945 all the Pacific countries became increasingly aware, with changes in communication, that they had many common interests, which made their scientific concerns more than simply academic. Wars and economic change, with rapid industrialisation in some areas obscured the potential unity, but the Pacific idea grew faster as the century moved on. In the late 1980s the Pacific Rim became an area of very fast economic growth, with a high potential for development in science and technology, holding promise of increasing scientific contributions to ICSU.

In several regions one saw countries which were not yet ready for formal full national membership, but had a part to play in, or benefits to be gained from, access to ICSU. For them the new membership category of National Scientific Associate was created in 1967.

A roll of new admissions is a reminder of the variety of national characters to which ICSU has to accommodate itself. Two members of the former British Empire adhered to ICSU, Sri Lanka in 1961, Jamaica in 1966. Turkey adhered in 1968 Iran in 1963 and Madagascar in 1970. The variety of cultures shows how ICSU has had to operate within a structure which had begun as European–American, and was now rapidly having the internationalism it originally claimed turned into a reality. Even the two Vietnams survived an extraordinary war, and the Vietnam which emerged retains its membership.

We have referred to ICSU's recognition and acceptance of its universal responsibility to the whole community of science of all nations by its creation of the Committee on Science and Technology in Developing Countries: COSTED. It was not an easy birth. The action of the 1966 General Assembly in Bombay had a long gestation, beginning at least as early as the Vienna General Assembly of 1963. Letters had been sent to Unions and to scientific contacts in many developing countries to get their reactions to proposals to form some system of aid, but those Unions which replied generally gave the impression that they already had their own individual methods of helping, and the response from developing countries themselves had been poor. However, when it came to serious discussion at Bombay, the Third World response was more encouraging, the most striking observation being that the stimulation of creative scientific work was more important than the charitable provision of food. This may be a contentious, even insupportable, opinion, but it does prevent one being too naive about the form aid should take.

It is hard to avoid using the word 'development'. It is the simplest word to label (if not always clearly to identify) the common elements in the economic and social changes which are a leading factor in recent history of what were disadvantaged communities. 'Development' has its weaknesses, but it does at least help us to bring together our thoughts on the common factors of movements of change for what is believed to be the better. The point was made by Professor Prawirohardjo at the Bombay meeting: development, he insisted, conveyed the idea of overcoming stagnation in countries to which change had to be brought.

The debate was an example of the convergence of minds from North and South. It also showed how scientists associated with the highest reaches of intellectual inquiry, at the Nobel Prize level, could assist in promoting the growth of scientific work from its base-lines within countries that needed it. P M S Blackett spoke to this theme from long conviction.

It also brought out the concern of the communist world for the advancement of the Third World. The proposal to create COSTED commanded the support of the USSR, a fact which suggests that all international politics is not necessarily adversarial.

During preliminary discussions with UNESCO, the question of aid for ICSU's own work in developing countries met with a chilly reaction. It was recommended that all ICSU should do was to get in touch with Education Ministries in needy countries who would argue within UNESCO for support for ICSU's work. This view did not persuade ICSU. A review was indeed made of the kind of heroic exercise that might be undertaken (such as training, creation of non-governmental scientific organisations within countries, direct advice on use of natural resources). These were all considered unrealistic. The best course would be to create the committee, COSTED, competent to take a long-term view of all the problems that might arise. There could then be a continuous creative survey of the problems of countries in the course of their hoped-for development.

Committees come and go: COSTED came and stayed until reorganised in 1993, when it was combined with the International Biosciences Networks.

Education

For a long time many people thought of the scientific community as academic and inward looking, often concerned mainly with the perpetuation of scientific scholarship, although many were also aware that there existed some professional organisations in some countries, limited on the whole to the efficient discharge of official duties of public or commercial employment.[10] *Science for the Citizen,*[11] to use the title of a very popular book which came out first in 1938, was one of a centuries-old line of books aimed at explaining science to the layman. Training for a scientific career is just as old, but after 1945 ICSU faced up to the need for an outlook which is more than popularisation, more than presenting science to a citizenry for whom science is something outside their lives. They looked also for something which is more than mere training in the skills of the individual scientist. They looked to education in science as something which would affect whole communities by improving the position and influence of the scientist as a citizen.

Individually the Unions shared in a widespread concern for education, taking this to mean not only training in the professions of its members, but also the spread of appreciation of the social value of its sciences. ICSU saw the larger scene and helped to establish world-wide studies of the need for education in science. In 1968, ICSU formed an Inter-Union Committee on Science Teaching (CTS). This body had a strong base. Several of the Unions had set up their own committees or commissions on the teaching of their own special subjects, and this CTS provided a means of co-ordinating their work. One aspect of it was of particular importance as ICSU's work for developing countries expanded: CTS was able to look beyond the needs for education in particular subjects to principles of study specially applicable to all of them. There is no single prescription for such work; no two developing countries have quite the same educational background to their basic science needs, nor the same educational structure on which to build. However, by having machinery for drawing many educational problems together, CTS was able to work on a broad front.

CTS was eventually one of the contributors to the UNESCO discussion which led, in April 1973, to the formation of ICASE, the International Council for Associations of Science Education. Why should there be two international bodies apparently doing the same thing? In fact, they are not. CTS began with science and scientists already organised on a world basis, and was concerned with developing science through education. ICASE began with a local study of the local needs of the teachers themselves, introducing science into general education. It began tentatively with a UNESCO Regional Workshop on integrated science teaching held in the Philippines in 1970. This led to a meeting of representatives of science teachers associations in Asia, the local organisation being provided by the Science Teachers Association of Singapore.

So condensed an account reads as if it were all chilly formality. Not a bit of it. The inaugural meeting was funded by UNESCO and ICSU, but the new organisation had not one penny to bless itself with to start even a newsletter to announce its birth. At the end of the meeting, Mr Jon Kusi-Achampong, who had done much to set up the Ghana Science Teachers' Association, passed round a plate to collect personal donations from the participants to provide a meagre but effective starting balance. $250 does not seem much, but it stands out among the tales of millions of dollars one has to tell, as proof of the personal commitment which so often matters more than the money.

It was in this spirit that ICASE flourished.

One world

Pause to distinguish between 'world-wide' and 'global'. It will have a bearing on the outlook of ICSU as the century progressed. We use both words rather casually, but it is useful to reflect on the very different meanings they can bear. 'World-wide' (at least, as it is used here) implies the involvement of some organisation in activities specific to its characteristic interests or scientific scope in all parts of the world. 'Global ' (as it is used here) signifies an interest in the whole planet, its several concentric spheres from core to outer space, the biosphere, and the geosphere. In the mid-century the word 'global' was not yet in use in titles of official organisations. It was to take a generation for this to happen, although the global idea was implicit in much that was being done.

A world view may originate in many ways. IUBS, for example, took a view of its own field of work, the world of biology, as boundless. Some Unions, like IUCr, limited their concern to particular subjects studied with particular techniques, but aimed to encourage these activities everywhere it could. Some looked to world-wide activity through devising membership systems which gathered in existing organisations, with common purposes throughout the world. Some, like IUGG, which could look back to an ancient tradition of 'earth measuring', saw the whole planet itself as the physical object of observation and experiment. IUPAC had long found subject-matter in an immense variety of materials, of human activities, and of means of identification, measurement, and conversion.

We live at the bottom of a shallow, restless layer of air. Ever since the 18th century we have been taking an interest in its local variations and their relation to local weather. These studies have grown into the modern analysis of the general weather and beyond even that now to the global atmospheric phenomenon. In 1967 a Global Atmosphere Research Programme was instituted jointly by ICSU and WMO. It was said of GARP's relation to ICSU and WMO that the two organisations not only represented non-governmental and governmental scientific structures, but between them could claim to command allegiance of the whole of the atmospheric science community. As described in a later chapter, GARP continued until it was eventually absorbed into the World Climate Research Programme (WCRP) (1984).

There thus appeared new bodies expressing the realisation that humankind was affected by the physics, the chemistry, and the biology of an objective world, all of which humankind might study. In the 1960s there was added a new, and, it could be said, more fundamental view: that humankind is itself an agent in change. The International Biological Programme (IBP) (which is dealt with in chapter 13) had to take this view of its selection of research topics. In 1969 ICSU saw the need to put the central objectives of IBP on its own permanent footing, and created a dedicated Scientific Committee on Problems of the Environment (SCOPE) of which a special commitment was the study of human influence on the environment. It took some years for effective programmes to be developed, so we can return to SCOPE's work in a later chapter. SCOPE was, however, the first international body to see the need to bring the great variety of humanity-related environmental problems together under the view of one organisation.

At mid-century, was the position of ICSU ideal? With regard to scientific scope, its links with engineering, agricultural, and medical sciences were weak and they have not improved since as much as many would like. More interaction existed with the Social Sciences, united since the Second World War in ISSC (International Social Science Council). Some ICSU bodies had interests which related to this body, so affiliation to it was of general benefit all round.

It seems rather obvious that IUPsyS, for example, should find common ground with other members of ISSC. Less obvious is it that IGU should do so, until one recalls that it was the geographers (as early as 1925)[12] who were the first to consider whether human population problems, particularly overpopulation, should be addressed by scientists engaged in physical and biological problems. We see even better evidence of the value to both Unions of affiliation with ISSC if we list some of their commissions. Among the components of IUPsyS one sees the International Network of the Young Child and Family and the International Network of Research Centres in Behavioral Ecology/Environmental Psychology. Are these so very far from the IGU Commissions on Health, Development, and Environment or on Population Geography? The connections suggest matters for inquiry in the future. A beginning was made in October 1993 with an international conference on population sponsored by 15 national academies.

Furthermore the IUHPS could be considered to be in the social science area, although it adhered only to ICSU. The International Union of Anthropological and Ethnological Sciences (IUAES) (1948), a member of ISSC, joined ICSU in 1993.

In the early 1990s, ISSC launched its Human Dimensions of Global Environmental Change Programme (HDP) in parallel with ICSU's IGBP, which, at the start, had decided to restrict itself to the natural sciences aspect, which was thought to be complex and difficult enough in itself. By 1994, co-operation between the two programmes was growing, but arrangements of formal sponsorship for such projects did not follow until 1995.

From this mix of different aspects of ICSU, its administration, its components, its creative functions, its world view, we can get a picture of an organisation any scientist, 'pure' or 'applied' might encounter anywhere in the world through any kind of work. 'Might' is the word to watch. None of the features set out in this chapter describes something the ordinary working scientifically trained man or women would be bound to meet. Most never hear of ICSU. This is not surprising. Most scientists are passive in their membership of their professional organisations. All the same, it is difficult to think of any of the main scientific occupations which does not in one way or other have some contact with ICSU, even if only through personal or corporate membership of some national or international professional association which is a member of an ICSU Union. But, as these pages show, the involvement of ICSU with growing science has grown with it.

But were these deficiencies in scientific breadth and image among scientists real drawbacks? Avoiding trying to be 'all things to all people' could also very well be a source of strength. However this may be, at mid-century ICSU was poised to embark on a series of interdisciplinary activities which led to its present strong role in the science of the global environment. The basis was in the International Geophysical Year (IGY), launched in 1958, one of the great events of the 20th century.

Chapter 12

The International Geophysical Year and its heritage

IGY: conception

Standing out in the 1950s is the International Geophysical Year (IGY), something we should think of as being a creative influence as well as an event. One of its long-term effects had been foreshadowed in what might be called a consolidation exercise: the formation in 1956 of FAGS: Federation of Astronomical and Geophysical Services. Several of the Unions had long provided useful services of gathering, interpreting and distributing data in their own fields, and were now persuaded to take part in a unification of this practice. ICSU and UNESCO[1] agreed to sponsor this Federation, which continues in existence. We have here an early example of the interdisciplinary policies of ICSU which have extended year by year ever since. The formation of FAGS owed a great deal to the personal skill and enthusiasm for ICSU affairs of G. Laclavère,[2] an outstanding example of the kind of person ICSU seems always able to attract to its work. FAGS turns up later, but its significance here is its outward-looking character: there were many reasons for the energy with which IGY was launched, but the trend towards the unification of functions which FAGS represented was one of the most important.

Three events make up a picture of expansion. The first International Polar Year (1882–3) was an affair, on the whole, of individuals backed by national academies. The second IPY (1932–3) was initiated and energised by the International Meteorological Organization, that is to say a body concerned with a limited subject. The third IPY, which became the International Geophysical Year, was the affair of an international body concerned with science at large, namely ICSU.

Eighteen months make a long year. That is how the organisers of the International Geophysical Year had to view matters when they looked back at the history of two International Polar Years and considered whether to try to organise another one.[3] The first had been little more than an actual year, from 1882–3. Its success was well remembered. Proposals to mount another Polar Year 50 years on were welcomed and good plans made. Unhappily the preparatory years and the years of actual activity were those of world economic perturbation, and the onset of the political and military disturbance which culminated in general war. All the same, much good came out of the Second International Polar Year (1932–3) and the appreciation of the two Polar Years together remained vivid in the minds of many who saw the chance of a much more fruitful exercise in the years not far away.

The hampering of the Second International Polar Year did not mean that it just stopped and lost all its value: far from it. An exploratory project like this does not stop when the last field worker has hung up his boots or the last instrument has been turned off or the last specimen numbered. Much of the work of analysis and interpretation only begins then. A great deal of benefit had resulted from the Second IPY, but by 1946 it seemed that there was no way of extracting further profit from its observations and records. However, when in July 1946 the International Meteorological Organization[4] formed a Commission for the Liquidation of the Second Polar Year, it appeared that, although it must wind up the interpretation of the accumulated results by December 1950, this need not be the end of everything. Much had changed, and much more was now possible.

The great advances in instrumental technique in many areas of science during and after the Second World War had encouraged many investigators to consider tackling problems which had formerly seemed beyond their reach. In spite of the aftermath of resentment, there was a stronger feeling than ever before of the possibility of international co-operation, and in spite of the obstacles of the so-called Cold War and the Korean conflict (1950–3) progress was made, especially after the ending of hostilities in Korea and the political consequences of the death of Stalin in 1953.

Big things often start small, with conversations among a few colleagues. On 5 April 1950 a number of scientists met at a private dinner at Silver Springs, Maryland, at the invitation of the American James van Allen[5] to honour the British Sydney Chapman[6] from Imperial College London. Their main common concern was current work on the Ionosphere, a subject which had been one of the concerns of the Second IPY but could now be studied more profoundly. They found themselves mutually interested in the idea of new ventures in the study of the upper atmosphere, of a kind which could well lead to yet another IPY. The path from an idea to an administrative reality in international science is nowhere better exemplified than in the way a '3rd IPY' idea became an 'IGY' in action.

Why was the Ionosphere so important? One of the great discoveries, theoretical and practical, of the 20th century had been the ability to communicate at far longer distances than might have been supposed if radio waves could only travel in straight lines. It had been found that they were in fact reflected or refracted by layers of ions high in the upper reaches of the atmosphere. Radio had begun by Herz's transmission of a signal a few metres, from point to point in a straight line. Now it circled the world, and the explanation for this unexpected behaviour had been found to lie in the upper atmosphere, in an ionosphere. The radio scientists were impelled to think globally, both in the realm of theory and as regards practical utility.

ICSU had formed a Mixed Commission on the Ionosphere, the sponsors being URSI, IAU, and IUGG, and in this forum the 3rd IPY idea was discussed further. In the summer of 1950 it came up again at conferences, at China Lake in California and at Pennsylvania State University, of this Commission. Marcel Nicolet,[7] who was to play an important part in the subsequent administration, was present at some of these meetings, and the understanding he developed of the issues involved proved to be a great source of strength to the project. The notion was well received whenever it was aired. In July 1950 there was a meeting of this URSI/IAU/IUGG Joint Commission on the Ionosphere in Brussels under the chairmanship of Sir Edward Appleton,[8] which adopted the following resolution:

> that the Third International Polar Year be nominated for 1957–58 and that, in view of the length of time necessary for adequate organization of the complex physical equipment now potentially available, an International Polar Year Commission be appointed in 1951 to supervise planning.

The resolution was accompanied by detailed argument in support.[9] It was put to the three Unions sponsoring the Commission and to the Bureau of ICSU. In September 1950 URSI endorsed it at its General Assembly and IAU at its Executive Committee. IUGG followed suit at a General Assembly in August 1951. Considering that most international committees are made up of busy men snatching time from other duties, this was quick work.

The proposal for a Third International Polar Year was put to the Executive Committee of ICSU in January 1951, and then to the ICSU Executive Board in Washington in October 1951. Of the possible types of body which might be appropriate, it was finally decided to appoint a flexible Special Committee, to be formed after consultation with, and approval by, the relevant Unions. The suggested date was to be 1957–8 corresponding with the next phase of maximum solar activity. The question was to be discussed in detail at the 1952 General Assembly of ICSU in Amsterdam.[10] In the event this programme was carried out somewhat later.

While this was going on there was another necessary actor on the scene: the World Meteorological Organization (established by UN in 1947). The collaboration of WMO was very necessary, as had been pointed out by the International Meteorological Association. With the International Association of Terrestrial Magnetism and Electricity, IMA had drawn attention to the importance of studies in their fields, which would be incomplete without reference to low latitudes. The ionosphere experts were interested in what went on in the high atmosphere. WMO and IMA were interested in what went on nearer and under the earth's surface. The Executive Committee of WMO agreed to participation, but pointed out that, since the scope of the studies envisaged went so far beyond matters of purely Polar interest, another title might be preferable. They proposed 'International Geophysical Year'.

From 'Polar' to 'Geophysical', from regional to global: the striking change of name was more than a trivial convenience. Science changes as fast as music or fashion or politics. By the time the new 'Polar' Year became a possibility, the sciences concerned with polar matters looked very different from what they had done in 1932. The repertoire of experiment and measurement had been extended in novel ways, both by new instrumentation and by new ways of integrating observation.

The two Polar regions were viewed now as integral to the world climate system, to the global magnetic structure, and to much else. The problems which faced the community of potential investigators were thus geophysical in the widest sense. In the view of WMO, 'Geophysical Year' would be a better term, and this was seriously considered by the Unions.

How long should the programme be? Many of the phenomena to be studied were cyclical or of such long duration that to terminate work sharply after a calendar year might truncate many useful surveys. The Antarctic was going to be as significant as the Arctic, so a full cycle of seasons at both Poles needed a year and a half to survey fully. In particular it was wished to relate to a full year's activity of the sun, which did not oblige by coinciding with a terrestrial calendar. A nominal central twelve-months period was agreed on, although preparation and analysis would make the effective life longer. (Later, in 1954 the period was extended: from July 1957 to August 1958, and later to December 1958.) The scope of investigations was far wider than had been possible only a generation previously. So the 'year', whatever it was called, was to stretch beyond a calendar year.

The constituent Unions of the Mixed Commission on the Ionosphere were unanimous in their support of the IGY idea, IUGG in particular being very enthusiastic. URSI went ahead with organising ionospheric and radio-science programmes. A Joint Commission for the Study of Solar and Terrestrial Relationships (on which URSI, IAU, and IUGG were represented) passed a resolution (3 September 1952) which still referred to the year as 'Polar'.[11] However the 'geophysical' idea was rapidly taking over, since it was now evident that the interests were in phenomena not only at all latitudes but in all components of the global system.

ICSU now took on the administrative lead; in October 1952 the Executive Board and General Assembly, in Amsterdam, confirmed the change of name and created the Comité Scientifique pour l' Année Géophysique: CSAGI. (For almost the last time the French language was used officially for a title. It survives now only in the euphonious URSI.)

In its new formal guise CSAGI, with its ICSU authority, was able to invite all the interested Unions to submit programmes of the kind of observation they considered should be made. As important, for several good reasons, were invitations to national members of ICSU. It was vital to the success of any such scheme to keep in mind that the scientific world coexists with the political world which creates, supports, sustains, or tolerates the national academies. National support could be effected through finance, through national research councils where they existed, and through national scientific establishments working in the scientific fields with which IGY would be concerned. There may be no physical boundaries to the world of learning, but nearly every scientist has to work with his feet on the ground of some nation, with that nation's support or consent.

The USSR was a problem. It was not a member of ICSU itself, but its scientists were active in many of the IGY fields of interest and there were many personal contacts with USSR scientists. The approach to it was therefore distinct, and, in the end, effective. Each of the participating bodies held its own meetings and gradually evolved policies towards the form and content of programmes. The meteorological programme was considered by a working group of the International Meteorological Association, and useful additions were proposed by A A Solotoukhine, the USSR representative in WMO and a member of its Executive Committee. This show of interest by the USSR was extended at the meeting of CSAGI in October 1954, which was notified that the Academy of Sciences of the USSR had accepted an invitation to take part in IGY, and had not only appointed a national committee, but undertook to encourage other nations to follow its example.

How do you make a 'world' programme, a programme in which those who want to take part can identify roles that will fit their resources and individual objectives? The General Assembly of IUGG in Rome, in September 1954, set about it by formulating criteria as follows:

A The programmes of the IGY should be selected with a view to solving specific plane-tary problems of the earth. To achieve such solutions it was recognised that during the IGY the regular scientific facilities of the world must be supplemented by additional observations suitably distributed, in space and time, as needed for the solution of selected problems.

B Problems requiring special attention during the IGY should be selected according to the following criteria:

  (a) Problems requiring concurrent synoptic observations at many points involving co-operative observations by many nations.

  (b) Problems of branches of geophysical sciences whose solutions will be aided by the availability of synoptic or other concentrated work during the IGY in other geo-physical sciences.

  (c) Observations of all major geophysical phenomena in relatively inaccessible regions of the Earth that can be occupied during the IGY because of the extra-ordinary effort during that interval, in order to augment our basic knowledge of the earth and of solar and other influences acting upon it.

  (d) Epochal observations of slowly varying terrestrial phenomena, to establish basic information for subsequent comparison at later epochs.

C Where questions of priority arise, priority should be given to programmes under sub-paragraph B(a).

Obviously one could not give the same attention to every area of the earth or every minute of the year. Some special attention had to be agreed under both heads. Geographically it was agreed to pay special attention to the Arctic and Antarctic polar caps and to three meridians, one running down the west coast of America, one through Europe and Africa, and one through Japan and Australia.

As regards special times, a system of World Days was set up. Observation was co-ordinated on certain days, chosen (a) at regular intervals, (b) on the advice of WMO, short periods at the solstices and equinoxes, mainly but not exclusively for meteorolog-ical observations, and (c) on alerts such as sudden displays of solar activity.

It was anticipated that the existing official and scientific radio network would be able to keep up the necessary traffic in communication. Collaboration at the same laboratory bench may be easy enough: collaboration at several thousand miles distance needs technical back-up of some complexity. The world radio network was by now competent to cope with IGY needs.

Preparatory phase

Many people and many bodies agreed to play a part in IGY, to provide accommodation, to involve apparatus. Credit must also be given where it is often forgotten: to those who agreed to release staff to work with IGY.

Help was needed also in the political/financial realm, for which ICSU, occupying a key position, approached UNESCO and other bodies on behalf of the CSAGI members. By 30 June 1953 a plenary meeting of CSAGI in Brussels was able to hear that 30 academies had agreed to participate. It elected the highly respected Sydney Chapman as President, Lloyd Berkner as Vice-President, and Marcel Nicolet as Permanent Secretary. The preparatory work had been done most efficiently by Col. E Herbays;[12] the Committee moved into action quickly. Through the courtesy of the Belgian Government, Nicolet was able to set up his office at the Royal Meteorological Institute of Belgium. For the time being funds came from ICSU.

Between IUGG, IAU, and URSI there were 14 separate identifiable disciplines or disciplinary areas represented in the programme of study. For each there was appointed a Reporter:

      I World Days (A H Shapley. USA)
     II Meteorology (J van Mieghem. Belgium)
    III Geomagnetism (V Laursen. Denmark)
     IV Aurora and Airglow (S Chapman UK, with F Roach and C Elvey USA)
      V Ionosphere (W J G Beynon. UK)
     VI Solar Activity (Y Ohman. Sweden)
    VII Cosmic Rays (J A Simpson. USA)
   VIII Longitudes and Latitudes (A Danjon. France)
     IX Glaciology (J M Wordie. UK)
      X Oceanography (G Laclavère. France)
     XI Rockets and Satellites (L V Berkner, USA)
    XII Seismology (V V Beloussoff. USSR)
   XIII Gravimetry (P Lejay. France)
    XIV Nuclear Radiation (M Nicolet. Belgium)

The organisation and committee arrangements were, as might be expected, more of a weave than a structure. The warp, one might say, was that of the Assistant Scientific Secretaries, who organised the many special conferences necessary. The woof was provided by the Unions and by ICSU systematising the practical work it was hoped to carry out. The forward planning included the preparation of 12 instruction manuals which were made available for wide study before the beginning of the period of observation. To the outsider this might all seem ponderously dull, but not a bit of it. There was a common determination to make this huge project succeed, and those who were close to the centre experienced one great thrill. The reader of these words written in the 1990s may find it difficult to grasp what it felt like to those attending the Rome Assembly in 1954 to hear a proposal from the USSR to use rocket-launched instruments as part of the experimental programme. They proved to be invaluable and to contribute new kinds of survey of atmospheric events.

Perhaps the most dramatic justification of the 'Geophysical' title occurred in the fourth month of the IGY: the launch of the first earth satellite, Sputnik I. Until now the world as a whole could be seen only in the imagination. Soon it could be surveyed with instruments from nearby space. It was not long before it could be seen by a human eye. These are dramatic elements in the story. The value of IGY lay in the solid scientific achievement in those 14 disciplinary areas.

Implementation and results

So, an enormous task was set in hand which could not have been carried out by people hampered by financial, diplomatic, and administrative restraints. Such restraints there were, but for the most part they were overcome. Doubts and difficulties came to the surface during the IGY planning period which needed all the personal qualities of leading personalities to resolve. The Vice-President of CSAGI (Berkner) was elected President of ICSU which, at some moments put him in an awkward position. He had to resolve in one capacity conflicts in which, in another capacity, he was party. Nicolet later described him as being 'a victim of the conflict between politics and science that still characterised the world situation at the time'.

The involvement of the USSR and of the Republic of China produced some tense situations, but men like Beloussov and Coulomb showed great skill in easing tension and making it possible for these nations to contribute. In the end political stress failed to hinder achievement. IGY became a model for international co-operation in the face of political stress. It even led to one political triumph, the Antarctic Treaty.

The last meeting of CSAGI was followed by the first meeting of an International Geophysical Committee which aimed to encourage continued international activity in geophysics. One of its first inspirations came from the exceptional solar activity experienced during IGY, and it was resolved to plan the International Years of the Quiet Sun.

IGY cost a lot of money in funds and money's worth in services.[13] To begin with, ICSU contributed out of its own funds and out of the subvention it received from UNESCO. Then UNESCO responded to an appeal for a direct contribution. Some national bodies helped, including the US National Academy of Sciences and the Royal Society. Later appeals were made to other nations for help, and this was forthcoming from over 18 nations in addition to further contributions from the USA, the USSR, and the UK.

One of the heaviest costs was in publication. Sir Harold Spencer-Jones was invited to become General Editor, and he set up an office at the Royal Institution in London, from which he produced instruction manuals and much other working documentation. He also began the process of negotiating with publishers over the eventual official publication of all the results of the programme. That series of *Annals of the IGY* was produced by a fairly new firm, Pergamon, which maintained its place in the scientific publishing field long after.[14]

Even so large a publication could contain only a fraction of the data which emerged. It was agreed to set up several World Data Centres to receive, catalogue, and communicate the detailed findings, each nation to meet the cost of sending its findings to a World Data Centre, the origin of which we take up again below.

The results of the IGY studies are too many and too detailed to list here, but one must make one remark of interest even to the wide non-scientific public. It is now clear that one of the purposes (even the main purpose) of expeditions and explorations in the remoter parts of the earth are now geophysical in intention.

IGY was the beginning of the new view of the earth that characterised scientific cooperation in the second half of this century.

Heritage

What, then, is the significance of IGY to ICSU today, 40 or more years after it was first suggested? IGY could not have existed without ICSU, but IGY changed ICSU. From now on ICSU could confidently plan to initiate programmes on a much larger scale than ever before. Why not say 'This is going back to the IRC outlook of telling Unions what to do'? Because the same considerations of freedom of action that ought to apply to individuals applied, throughout the planning of IGY, to Unions and to National Members. There was no pressure in the IGY negotiations, but the freedom of communication between colleagues played its part in bringing about participation by nations which might otherwise have held aloof. The pattern achieved was not perfect, but with 67 nations involved we can say that it ended up good enough to nourish hope for the future.

There exists one certain proof that widespread scientific action can have political consequences of extent and value: the Antarctic Treaty. The work in the Antarctic on many of the IGY topics exposed national rivalries in the ill-defined territories. These were, on the whole, resolved harmoniously. Bases were established by a number of nations, 11 of them having become permanent. The resolution of the political tensions, which had been exposed during the long debates over IGY action in the region, resulted in the formulation of an Antarctic Treaty which came into force in 1961.

But it is the scientific data which matter. The massive results remain on record and are of permanent value, but just as important is the precedent IGY set for collective enterprise in scientific work on a large scale. Because the IGY system had succeeded so well, it was possible to view other areas of scientific study as possible candidates for the same kind of wide-ranging organisation on a long-term scale. Directly out of the physical science areas of interest of IGY came such bodies as COSPAR (Committee on Space Research) and SCAR (Scientific Committee on Antarctic Research). COSPAR was set up in 1958 when it was seen, during IGY, how important research in space, and from space, was going to be. The membership style matched that of ICSU itself: national academies and Unions. It has continued to grow in the variety of its interests.

SCAR (Scientific Committee for Antarctic Research) was formed at the same time with rather more specific responsibilities: to maintain the impetus of Antarctic research, and through doing so to act as adviser to the signatories of the Antarctic Treaty.

This account may give the impression that during its life IGY was the only thing that mattered in wide-ranging science. This is far from true. ICSU was identifying other problems which needed just such a global view. Oceanic research was one such, and in 1957, following a meeting in Göteborg, ICSU created SCOR, Scientific Committee on Oceanic Research. SCOR took the world's oceans as its province, although concentrating on some areas of special interest such as the Indian Ocean and its potential as a food source. SCOR's operations were the sum of many smaller parts: for many years it worked through small groups working on problems which offered prospect of solution in a relatively short time. The sum of these efforts was considerable. (In recent times SCOR has embraced larger problems to which we shall return.)

There was more to follow. The IGY had consolidated the vantage-point from which science interacts in the globe as a whole. This is seen first, in IBP, in SCAR, SCOR, COSPAR, and the WDCs. It later prepared the ground for ICSU's work in the global environment through SCOPE, WCRP, IGBP, the Global Observing Systems, and GARP.

GARP, the Global Atmospheric Research Programme, is the first international research programme carried out jointly by ICSU and a UN organisation. It is therefore worth while taking a closer look at its genesis.

On 25 September 1961 President J F Kennedy, in a major speech before the 16th session of the General Assembly of the United Nations, referred (*inter alia*) to the peaceful uses of outer space. He said:

> we shall urge proposals extending the UN Charter to the limits of man's exploration in the universe, reserving outer space for peaceful use, prohibiting weapons of mass destruction in space or on celestial bodies, and opening the mysteries and benefits of space to every nation. We shall propose further co-operative efforts between all nations in weather prediction and eventually weather control. We shall propose, finally, a global system of communication satellites linking the whole world in telegraph, telephone, radio and television. The day need not be far distant when such a system will televise the proceedings of this body in every corner of the world for the benefit of peace.

On 4 December 1961, Adlai Stevenson, on behalf of the USA, introduced a draft resolution to the General Assembly First committee which called for 'a world wide effort in weather research and prediction under the auspices of the United Nations'.

After debate and revision this was sponsored by 24 nations. Its paragraph 2 reads: '[The General Assembly] requests the World Meteorological Organization, consulting as appropriate with UNESCO and other specialized agencies and governmental and non-governmental organizations, such as the International Council of Scientific Unions, to submit a report . . . regarding appropriate organizational and financial arrangements to achieve these ends].' The final text was adopted in the Plenary Session of the General Assembly on 13 December 1961.

The outcome of these consultations as laid down in the report led to Resolution 1802(XVII) which was adopted in the Seventeenth Session of the General Assembly (1962) and (*inter alia*) 'Recommends that the World Meteorological Organization, in consultation with other UN agencies and governmental and non-governmental organizations, should develop in greater detail its plan for an expanded programme to strengthen meteorological services and research, placing particular emphasis on the use of meteorological satellites and on the expansion of training and educational opportunities in these fields' (paragraph III, 3).

Paragraph II, 4 reads: '[UN] invites the International Council of Scientific Unions through its member unions and national academies to develop an expanded programme of atmospheric science research which will complement the programmes fostered by the World Meteorological Organization'.

Thus for the first time the General Assembly of the UN invited ICSU to join forces with a UN body in world-wide research. After some years this led to the joint programme GARP. No doubt the achievements of IGY had played an important role in this important step forward.

The single programme which followed most closely the organisational pattern of IGY was the IQSY (International Years of the Quiet Sun). In this there was co-operation from many organisations of many nations. There were many projects in it, all related to some extent to solar activity. Some were traditional, such as observation of an eclipse. Some were strikingly new, such as the use of space vehicles for observation above the earth's atmosphere. IQSY is now the achievement of a past generation, but many of its detailed observations and results remain valuable in current research. This has to be said of many of the programmes in which ICSU has been concerned: the collaborative efforts are cumulative in their effects on new science.

While the various follow-up projects of the IGY, in which ICSU was concerned, were unfolding, UNESCO was pioneering with other environmental programmes. After an abortive attempt to establish, in the Amazon basin, an international research station on humid tropical forest problems, it did organise an interdisciplinary programme of research on arid lands problems. The subsequent meetings and publications on arid zone research became the forerunner of a series of efforts to address such issues on a world-wide basis.[15]

Science could never be the same again after the period around IGY. Those involved had this in common: they all looked at their special interests in the setting of the world as a whole. This was destined to lead to a better understanding of environmental problems and to show that research in many areas is no longer to be limited to the confines of our living space on our own planet. Which does not mean to say that our relation with our immediate living space is of little account: a contrary view was the inspiration of another great enterprise. The biologists watched IGY with interest and set about developing their own world survey. IBP (the International Biological Programme) was the result, but its path and progress did not eventually copy the style or manner of IGY. The biologists did their own thing in their own way.

Chapter 13
Data, and scientific information

Data

IGY produced many results in the form of matter for study: it also stimulated changes in manner. For many years there had existed centres for the collection of data relating to all manner of activities, known to and used by specialists in limited areas, but not, on the whole, co-ordinated. IGY created a new attitude. The many research programmes, it was anticipated, would produce a vast amount of material which would need to be held in store and made available when required. The future need was met by the adaptation of Data Centres which already existed in some countries by putting these national and regional centres on an international footing, according them a common identity as World Data Centres.[1]

An old and important problem in the evaluation, storage, and retrieval of scientific information is how and in what form to make it accessible. This applies above all to numerical and statistical data. If an individual makes a measurement of permanent usefulness to others, how is it to be made known to them? In order to promote international co-operation and effort in this field, ICSU established a Committee on Data for Science and Technology, or CODATA (1966).[2] Like many a proposal which can be reported baldly as a committee resolution, this had long roots.

As we have said, scientists used to communicate by letter. Then they created journals. Two hundred years ago, there were perhaps 50. One hundred years ago one could reckon about 500. Seventy years ago there were 5,000. Today there are in excess of 50,000. It has been suggested that more than 90% of all scientists who have ever lived are alive today (a figure which depends to some extent on what you mean by a scientist).[3]

In order to help scientists find their way in the ever-growing complex scientific literature, many abstracting services like *Chemical Abstracts* and *Biological Abstracts* emerged, successors to the earlier *Catalogue of Scientific Literature*, but operating eventually on a comercial basis. To help create periodical overviews of the state of the art, numerous *Annual Review* type publications appeared and, in order to speed up access, publications like *Current Contents* appeared, announcing forthcoming articles by title and author. Superimposed on this, literature search and retrieval systems by computer became available to most scientists.

There may be more scientists, but the working time available to any one individual has not increased. Each individual is faced with the problem of the selection and co-ordination of information relevant to his work. He cannot deal with everything he might need. As often as not he does not know what it is he needs until he needs it. There have been many efforts at dealing with this problem, the chief field for such enterprise being in numerical data.

Results of particular investigations have always been published for the benefit of colleagues, but in the past century there have been many efforts at gathering scattered results together and publishing them as lists of general utility. In 1883 Hans Heinrich Landolt and Richard Börnstein published their *Physikalische-chemisch Tabellen*, running to 281 pages. It grew and survived: during the period 1950–69, 26 volumes appeared running to 20,000 pages. A French effort was the *Tables Annuelles de Constantes et Données Numériques* 1910–30. Some instalments even appeared during the troubled times 1936–45. It acquired a new title: *Tables de Constantes Selectionnées*. A British production was taken under the wing of the National Physical Laboratory. Kaye and Laby's *Tables of Physical and Chemical Constants* appeared in single volume form from 1911 to 1966.

In 1919 IUPAC embarked on a scheme of *International Critical Tables*. IRC gave the scheme its blessing in 1923; the National Academy of Sciences of the USA took on financial and editorial responsibility. It came out between 1926 and 1933 as a single edition of 7 volumes, 3,819 pages. The editor in chief, appointed in 1920, was Edward W Washburn. He died in 1934; the work died with him. In 1958 the US National Academy of Sciences established an Office of Critical Tables, of which the Director was Dr Guy Waddington. His knowledge and experience were to be valuable in another connection later on: the formation of CODATA.

IGY and the WDCs

The most long-lasting system we can identify as providing a co-ordination of services is one we have already mentioned in connection with IGY: FAGS (the Federation of Astronomical and Geophysical Services), set up in 1956 to bring together services, each of international value but to some extent isolated in particular countries and institutions, or relying on the support of particular Unions. The world-wide monetary inflation had reduced Union support for some ten services (ranging from that of the International Time Bureau to that of the World Mean Sea Level Service). ICSU discussed this with Unions and with UNESCO, arriving at a scheme which aimed at identifying common problems and, where possible, finding common solutions. Financial aid came from the countries of domicile of each of the services, but this could never be taken for granted: ensuring its continuance was a major pre-occupation. In addition to national support, the distribution of aid from UNESCO was seen to hold hope for the future. It hardly needs saying that none of this was easy. IGY put a heavy load on all available resources: redistribution by ICSU of its own resources for this kind of work was criticised.[4] What is important to us at this point is to see the beginning of a style of ICSU work which was to develop.

The organisers of IGY were nothing if not far-sighted. They decided, in 1955 at a meeting of CSAGI (the Comité Scientifique pour l'Année Géophysique) that traditional methods of reporting were not good enough, that something better than station books and expedition reports were needed if results were to be of long-term and widespread value. They decided that IGY data collections, both from monitoring and special data sets, should be preserved in special long-term data centres for future use. Moreover they suggested that efforts be made to handle the data in 'machine-readable' form. At the time this had to mean some punched card system, the modern electronic systems being still some way in the future. But confidence in the philosophy paid off. Out of this resolve came the idea of World Data Centres, each of which would establish an archive of data in a suitably accessible form. For example WMO became one of the WDCs for Meteorology, and published its collection in a microcard form. Other WDCs were established quite soon, and ICSU saw a need to co-ordinate the way data were handled, not only WDC data, but data deriving elsewhere, intended for other purposes.

Countries were invited to establish and fund new Data Centres, for the several IGY disciplines, and to ensure that the archive systems set up would have long operational lives. The first nations to volunteer were the USA and the USSR (creating WDC-A and WDC-B respectively). These covered all the main IGY disciplines. Some European countries, then Japan and Australia followed suit for additional disciplines. The World Meteorological Organisation, a UN body, then volunteered to be a WDC for meteorology. These latter were grouped together as WDC-C.

This was intended to be an adjunct to the IGY organisation, but after the IGY period ended it was clear that the WDC system was too good to allow to remain static. Some IGY WDC centres were allowed to subside, their holdings of data being transferred to other centres, but ICSU proceeded to create a Panel on World Data Centres, so making the system a factor in its own conduct of global geophysical programmes.

As ICSU thus enlarged its global and environmental commitment beyond the scope of IGY, it saw the concomitant need to enlarge its view of the range and utility of WDCs, so new ones were created. With this extension came also the involvement of other countries than those already active, notably China.

The IGY WDCs had been passive in their role, concentrating on collection and storing data, but not initiating any processing, which was to be left to the users in the course of their scientific enquiries. The WDCs, that is to say, dealt only with exchanging data, accepting data from authorised co-operating ICSU bodies, providing copies of data at copying cost, permitting and aiding access to data stores, publishing catalogues, helping to find data not known to be in the WDC system.

The WDC system came into existence before the rapid expansion in computerised data handling which characterised the 1970s and 1980s. The first changes which were brought about were in storage and transfer, but the interpretation and analysis of data which became possible also offered the chance to relate one collection to another and one discipline to another. (To anticipate: with modern on-line communications systems, world browsing of catalogues has become a possibility. New storage systems like CD-ROM have made the simple transfer of large amounts of material possible, so that, although the WDC remains a centre, it need not remain a closed unit.) The catalogue concept of the earlier data collections has been supplemented by the use of database systems which greatly enlarge the variety of searches that can be made. Nothing has been lost: the old material still remains accessible on film and on paper, for those who need it in the older forms. Forty-four separate discipline WDCs now exist.[5] The creation of the WDCs met one need and highlighted another: a need for some sort of mechanism for the critical organisation of data on a systematic basis, a task for which the WDCs were neither yet fitted nor organised. Machinery to meet that need (CODATA) was to be created by the fruitful General Assembly of 1966 in Bombay.

CODATA

Publish or perish: any organisation of learned and professional people has to publish just as a human being has to breathe. The need to publish in a suitable way is often one of the things that persuades people that they must get together. IUCr is a typical example of an organisation which made publication one of its first needs and duties[6] (and incidentally a solid source of income), and most of the other Unions, pre-war and post-war, put publication as one of their prime duties. In some cases, satisfactory journals already existed, long before any relevant Union was formed, and there were many bodies, like national societies, which ran excellent and influential journals. The ICSU Unions could not, and would not wish to, claim to be the begetters of all new journals, but they had an influence, since the creation of new journals was a matter for international consultation, and many publications of a new kind owe their origin to consultation in the committees of Unions. ICSU can therefore be seen as playing some part in a world network of publication, in science pure and applied.

But gradually the old arrangements of publication by learned societies or even commercial publishers, in the forms which had been familiar for centuries, were proving inadequate. What was needed was not necessarily new mechanical forms of publication, but a new attitude to what the working scientist needed to have accessible.

There is a change to be seen in the development of post war publication: an emphasis on data dissemination. IUCr, IUPAP, IGU all insisted on the importance of this, and therefore helped to create the climate in which the Scientific Committee on Data for Science and Technology (CODATA) was created.

In the mid 1960s Harrison Brown,[7] later an ICSU President, called attention to the lack of any successor to the *International Critical Tables*, and was asked by ICSU to review the position, together with a working group from six nations. His group's report to the Bombay General Assembly (1966) proposed that there was every hope that an international committee could carry out such work of continuous critical systematisation. As we have seen, there was already a history of the successful compilation of tables of data and constants, but these were one-off efforts. How was continuity to be introduced? Brown's recommendation was to implement a plan for seeking, examining, evaluating, and making available in suitable ways all published quantitatively expressed values of units, properties, and functions.

What was now proposed was of a much more interventionist nature than anything conceived hitherto, something so substantial as to require a permanent secretariat and world-wide communications. The General Assembly agreed to the creation of CODATA which should be concerned with all types of quantitative data resulting from experimental measurements or observations in the physical, biological, geological, and astronomical sciences.

The philosophy behind CODATA was different from that behind the WDCs. It was concerned with data far more various than that arising from the limited range of IGY disciplines. The WDC material was mainly numerical or statistical. CODATA embraced biological data as well, much of which could not be reduced to numerical terms. At first, WDC material of a particular kind was intended to be used by specialists in the particular field in which the data had been determined, although later this view was broadened. The founders of CODATA hoped and expected that data would be made accessible and useful to workers in fields different from the discipline of origin. 'The scientific inter- and cross-disciplinary nature of CODATA, in the use of data, dictates its thoughts and activities.'[8]

It was clear that what the United States National Academy of Sciences had been doing on a national level was in many ways a model of what could be done on a world level. NAS offered CODATA space in its new building in Washington DC for two years, and the services, half time, of an Executive Director, Dr Guy Waddington. Other staff came from the Institut für Dokumentation, Frankfurt, the American Chemical Society, and the International Atomic Energy Agency.[9]

The first objective was the preparation of an 'International Compendium of Numerical Data Projects', which benefited from the Office of Critical Tables of NAS putting its files at the disposal of CODATA. Much travel and world-wide inquiry of every Union and in every nation with a visible interest in data management enabled the CODATA office to complete this first task by the end of 1969. By 1970, CODATA had stirred up a very wide interest in the assembly and communication of data. One of its most influential efforts had been to establish a Task Group on Computer Use. The Secretary of this Group, R Norman Jones, has written 'nostalgically', as he says, of its early days. The disciplines represented in the Group were computer science, documentation, physics, and spectroscopy. They, and others invited to a conference of wider scope, confronted each other as strangers, wondering what sort of meeting they had been asked to. Gradually those attending successive conferences began to see, and to be enthusiastic about, the prospects for collaborative work on data management. In Jones' opinion a turning-point was a meeting at Freiburg in 1973. There were 101 participants from 13 countries and 4 international organisations. 'It is reasonable to say that this conference consummated the marriage between computer technologists and the scientific community.'

An organisation like CODATA lives by mutual understanding between officers, elected or appointed. It must be for someone else to enquire into the possibility that the solid progress made by CODATA when Edgar F Westrum was Secretary General and Nicholas Kurti was Treasurer was due in part to their both being excellent cooks.

As with many organisations, the CODATA secretariat began with temporary appointments, but in 1974 Westrum, and Kurti, with the help of Jones, put the staffing on a permanent basis. Bertrand Dreyfus, a French physicist was the first Executive Secretary. He died, much regretted, in 1979.

Phyllis Glaeser succeeded him, and was responsible for overseeing the extension of CODATA's interests into the biological sciences which was initiated by Masao Kotani.

Those who developed CODATA appreciated that data are common tools, that what is noted or recorded in one worker's experiment in one discipline may well be seen to be relevant to someone working in another discipline. The distribution of data is now extensive and takes many forms, of which one of the most striking is perhaps that of diskettes which can be used in personal computers by individual workers: the growth of CODATA to world activity has brought it round to an unprecedented service to the individual.

Its principles were accepted implicitly in its early days: they were made explicit in official publications as those of 'improving the quality, reliability, management and accessibility of data of importance in all fields of science and technology'.

In the early 1990s new problems became apparent: an avalanche of data had become available from satellite observations and many of these were not (yet) relevant to ongoing research, and, as was said, answered questions that had not yet been asked. Trying to extract required data from this vast and rapidly growing reservoir was like 'trying to drink from a fully turned on fire hydrant'.

A second problem was the commercialisation of data which tended to limit their accessibility to scientists. Thirdly, some governments put restrictions on data gathered by government services for reasons of national interest, including defence, which after prolonged negotiation often proved to be unnecessary.

Recognising these problems, ICSU asked CODATA in 1993 to keep a close watch on them and to advise the Executive Board on appropriate action. CODATA set up a Working Group on Access to Data to act as a watch-dog.

ICSU AB and ICSTI

We cannot trace here the ups-and-downs of every part of ICSU, but many of them would illustrate the interplay of identification of need, errors of judgement, sensible resolution of difficulties, and the change of one kind of structure into another with times and conditions. We can take one example here: the growth, decline and fall of what was planned to be an important organisation (the ICSU Abstracting Board, ICSU AB) and the halting growth of its successor (the International Council for Scientific and Technical Information (ICSTI).[10]

Long before IRC, ICSU, the WDCs, and CODATA, one ofter the other, became concerned with the mechanisms of storage and transmission of data, other organisations were concerned with how the scientist could get to know what had been written without having to read every published word himself. The published scientific litera- ture grew and grew. So there came into existence abstracting services, and Abstracting Journals, the significance of which can be seen if we look briefly at the scientific liter- ature at large.

A generation ago, in 1959, G-A Boutry wrote in *ICSU Review* an article which looked back on the previous decade and forward to new developments. He offered his readers a rough and ready distinction between scientific *fact*, scientific *information*, and scientific *news*.[11]. His descriptions help us to follow some important developments in ICSU actions. According to Boutry, scientific *information* may be present in the state- ments made by a scientific worker or a group of scientific workers. A scientific *fact* emerging from it can only be considered as established when several other workers, operating independently from the first, have been able to reproduce it from the direc- tions given in the first worker's paper. (Even then, new work may lead to reconsidera- tion.) If someone publishes an assertion unsupported by the experimental or observational evidence, it may be interesting, but has to be looked on as scientific *news*. This may have a certain value, but it is that of news of any other kind of event in a news- paper and subject to the same kind of judgement.

The production of scientific information had developed from the 17th-century network of personal letters through the serial publications of the newly emerging Academies to the variety of outlets of the present day. Boutry's 1959 survey listed these forms:

1 Journals or serials published by universities, government departments, private indus- trial groups, or laboratories, all of which publish papers written by the staff of the research departments they maintain.

2 Publications which Boutry describes as *National Journals*. These are journals published by scientific societies maintained by, or constituted of, men and women of science belonging for the effective majority to one nation. Such bodies are national academies or institutes based in one country, such as the American Institute of Physics. In this class are journals, few in Boutry's day but more numerous now, published by commer- cial publishers.

3 *International Specialized Journals*: publications devoted to a single branch of science, sponsored, managed, or recommended by a committee of scientists of several nations. Even in Boutry's day it was necessary to remark that the number of such journals was rapidly increasing (to a greater extent as time went by through the involvement of com- mercial publishers).

4 *Scientific 'Newspapers'*: these may contain brief reports of new discoveries (often pub- lished as 'Letters to the Editor'), review articles, accounts of meetings and congresses. At one time there were few, but now there are many more, directed at different levels of scientific sophistication. Some, like *Nature*, are the preferred medium for the first announcement of discoveries of great significance.

This is, of course, only a rough and incomplete classification, but it indicates the variety of printed scientific publication.

Here we must emphasise *printed*. Since Boutry wrote his article there have been important additions and changes to the physical media of communication. However, the first steps in the international organisation of awareness were taken with the printed word alone.

Another new element to appear in the last few decades is the practice of judging the performance of a scientist on the basis of the number of his publications. Employers rated his/her value higher if these were published in international 'peer review' journals and if they were cited frequently (as can now be determined with ease by the *Science Citation Index*).[12]

For many years it had been the custom in many journals to preface an article with a short abstract of its essentials. There had also grown up a number of publications which collected such abstracts for republication or, more often, prepared their own abstracts from as great a variety of journals as possible in a particular field to provide workers world-wide with an indication of what was being done, or, when the journals were accu-mulated in libraries, had been done in the past. Ten years before Boutry's article, in 1949, the Department of Exact and Natural Sciences of UNESCO, of which Pierre Auger[13] was Director, convened a conference to consider the problems of scientific abstracting. Attendance was mixed and various: 24 countries, 8 of the ICSU Unions. They talked for 6 days about the future of abstracting. (It might have been supposed that the chemists, who could boast the most highly organised abstracting service (*Chemical Abstracts*)[14] would have had most to say, but it was, in fact, the physicists who were most active. It seems that they were more conscious than others of a lack of resource which might now be filled.)

The requirements of a good synopsis had been spelled out by the Royal Society,[15] but the physicists wanted more: an internationally organised journal providing a service of abstracts for 'physics, both pure and applied, including astrophysics and the geophysical sciences, and for such branches of engineering as it may be appropriate to include'. These words were used in the final recommendation. The recommendations accepted what was common at the time: a bilingual system with English and French on equal footing.

The recommendations were accepted 'in the mist of the happy atmosphere which always surrounds counsels of perfection'.[16] A further meeting, mainly of physicists in Paris, soon afterwards suggested the formation of a Joint Commission on Physics Abstracting, to be formed by ICSU, which approved the idea at a General Assembly in Copenhagen in September 1949.

This Commission had its first meeting in Paris in December 1949, and very soon doubts came to the surface: the previous meetings, and this one, had included only users of published abstracts (librarians and researchers). The organisers had forgotten the people at the workplace: the compilers of abstract journals. A new search for support amongst the makers of abstracts and of abstract policy (the Royal Society, the Institute of Electrical Engineers, etc.), was fruitful: a conference of 'Editors and users of Abstracts' met in London in September 1950. Boutry explains: 'In 1949 a gathering of physicists explained to the Editors how they should go about their business. A year later the same kind of people were asking the Editors "What can we do to help you increase the efficiency of your journals?"'

Further discussion led to a proposal in July 1951 by the Joint Commission itself that it should be dissolved and replaced by a Board, on which would serve representatives of ICSU and of the Abstracting Journals. In May 1952, UNESCO and ICSU established an Abstracting Board, to begin operation on 1 June 1952. Although closely linked with ICSU, it was to have a legal personality of its own, established in Belgium (on 3 November 1953).

The Board had two main arms, the ICSU representatives and those of the Member Journals. Gradually ICSU-AB extended its field of influence, solving problems of coverage and speed of publication, as well as constantly concerning itself with the quality and effectiveness of the scientific literature. It moved into chemistry in 1959, and in 1961 it moved into biology with the admission of several biology abstracting journals.

The function of the ICSU Abstracting Board had originally been to make accessible clues or pointers to sources of information in the original descriptions by investigators of the results of their work. Other interests grew. Although its brief was a concern for abstracting and abstracts, detailed studies, including statistical surveys, raised other more general matters, such as the standardisation of abbreviations of titles, transliteration of Cyrillic characters, indexing, classification systems, and, increasingly, the challenges of electronic devices and their increasing interest as means of publishing and communication. The constitution of ICSU-AB (provider and user working together) looked effective and impressive, but only before it was put to the test of practice.

Its world view commended itself to UNESCO, which undertook with ICSU a study of the feasibility of a world science information system. This system became known as UNISIST, which developed the inevitable committee structure. A Central Committee looked into classification systems and the feasibility of a machine-readable form of journal title. This latter is an indication of the pressures which were to be felt by ICSU-AB to adjust to changing conditions.

As a component of UNISIST, ICSU-AB had some valuable achievements to its credit: an International Serials Catalogue, an International Classification Scheme for Physics, prepared in collaboration with IUPAP, the European Physical Society, and the Institute of Physics (1975,1978).

Eventually it was seen that the rate of growth of scientific information was getting beyond any system of collation based on traditional methods. There were practical problems arising from changes in the personnel of the ICSU-AB Secretariat, in its location (it had moved to the Hotel de Noailles in 1978) and, as always in these matters, financial stress. Communication with ICSU and with the Unions deteriorated. Input from the Unions fell off, and the resignation of the United Kingdom and the United States[17] was felt as a serious blow. It was claimed by some that the apparent lack of interest of working scientists in ICSU-AB was due to its very success in improving the flow of information, but, be that as it may, interest flagged. The merits of ICSU-AB and its claim to recognition were argued at several General Assemblies of ICSU, but it was eventually agreed in 1983 that 'the formal connection between ICSU and the ICSU-AB be terminated and that the status of ICSU-AB as a Permanent Service of ICSU also be terminated'. An associated recommendation was accepted, that ICSU reconsider its relations with ICSU-AB or its successor body when ICSU-AB had sorted itself out.

An interim scheme with the title 'Pegasus' was tried out, but did not have to take wing for long. A successor to the ICSU-AB was established in 1984 to 'increase access-ibility to and awareness of scientific and technical information'. ICSU-AB had oper-ated with a static outlook, namely that the information was there in print to be sought and found, improvement in the methods of finding being the major objective. Three Task Groups were set up which had, as a common characteristic, concentration on technical matters and means. At an extraordinary General Assembly of ICSU-AB in 1984 new Statutes and membership structure were adopted and a successor to ICSU-AB brought into existence: to be called the International Council for Scientific and Technical Information (ICSTI).

It saw the information problem as dynamic, one of movement of information to where it was wanted rather than of search by those who wanted it. It therefore identi-fied problems in information transfer, which were already of great variety and clearly likely to become even more so. One piece of terminology makes this clear. A Technical Planning and Steering Committee was replaced by a much more positive Technical Activities Co-ordinating Committee. Task Forces were set up to oversee activities in several kinds of need, some related to activities (Economic Issues, Electronic Publishing, Legal Aspects of Information Transfer), some to fields of study (Chemistry, Life Sciences), some to method (Interdisciplinary Searching). Some work could be done in conjunction with other bodies (the Multilingual Thesaurus in conjunction with IUGS). All were concerned more with what was practicable rather than merely what was desirable.

After 1985, ICSTI built up a fresh set of relationships with its own member bodies and others. With ICSU its connection is now as a Scientific Associate, a relationship which is of the greatest importance in consolidating ICSTI's links with the scientific community. That consolidation has required continuous effort, hampered by what, to the outsider, could look like much talk with few deeds. This may be unfair, but ICSTI's own accounts of its proceedings too often contain such expression as 'This project has not been pursued.' It had to admit that during the period 1983–9 'there had been a lag between the objectives of its programme of activities and the achievements'.

CODATA and ICSTI are two sides of a coin: one is concerned with material of research and the other with means of getting the interpretation of it to the right user. Ways of working together have become a necessary preoccupation of both bodies, but too many changes have been taking place in the character of the information transfer chain for ICSTI to maintain a rigid character for long.

Who or what is most dominant in scientific publishing? At the time of writing it is in a state of change with the introduction of new techniques for both recording and communication. ICSTI will have much to watch and to do.

Experimental innovation

There is another way of looking at the pressures which led to the formation of CODATA and ICSTI and their reaction to changes in them. It is easy, and right, to point to the importance of the computer as a prime aid to scientific work, so long as we remember that on its own it is as dead as a wooden abacus. Obviously, work like designing a new aircraft component or a new drug molecule has to rely on some selected input. While it is probably true that the computer has stimulated the development of instruments which it would have been impossible to design or operate previously, instrumental ingenuity is as old as experimental science. Before the computer came into prominence there was no slackening in the pace of innovation in instrumentation, nor has there been since. All manner of new types of apparatus were always being devised. In the simplest terms, some were for measuring, like the mass spectrometer; some were for seeing, like the electron microscope. The radio telescope behaved as if it could see, which had been true for two generations of the X-ray spectrometers which had revealed molecular structures.

Many of them had originated as unique research instruments, and then proved so valuable or versatile that they were adapted for routine use as standard laboratory equipment for research and, even further, for use in industry for investigation and control. One result of this was that many scientists found themselves so drawn into communication with others working in the same new ways that they found it worthwhile to form new associations for sharing of results, methods, and ideas. Some association took place within existing Unions, some outside. New science makes new scientists; new scientists make new groups; new groups share data. Any new means of communicating data will quickly be utilised.

The new methods of utilising data and information which were the common self-imposed duty of such groups were, on the whole, synoptic in character: they relied for their value on bringing together in a controlled way separate observations which could be collated to give a picture or map of an area of knowledge. In the case of the earth itself this was soon no longer a simile; it became true of the look of the earth itself and of the condition of the oceans and of the atmosphere. From the time of IGY, that is to say, a new possibility was available, namely the observation of the earth by satellite, which gave a direct view of certain conditions which hitherto could only be pictorialised by the accumulation of individual observations.

This is ironic: the improvement in scientific satellite facilities went hand in hand with improved military inventiveness and with the pressure of commercial desires. It is hard to separate the components. Be that as it may, the effect on many of the member bodies of ICSU was to bring within their grasp what had been almost dream-like ambitions to their predecessors.

We have looked at some events of the 1950s. A generation later, by the time the Ringberg meeting of 1985 was being organised, it was expected that anyone invited would be able to get there in a short time. Between 1955 and 1985 world air-transport had reached the level of rapid communication between all capital cities, most larger cities, and centres of population. If it were necessary for people to talk face-to-face it could be arranged quickly and the event carried out quickly. This intensification of business was world-wide, so that we are now in the position that the speed of reaching agreement is that of the human will, not that of physical movement. This applied not only to the scientific community, but to everyone involved in occupations above a certain level of sophistication.

To speculate on the future for a few lines before coming back to the 1960s: while people involved in industry, commerce, government, or military occupations could all meet their counterparts from other areas or other countries with ease, means were being developed whereby it was hoped people could confer with ease, as if around a table, but in fact separated by thousands of miles.

One new instrument is the telefax which now plays a valuable role in scientific and organisational communication, superseding telex, telegraph, and ordinary mail services. It is now widely used for preparation of multi-authored documents, for getting the written views of the members of international committees, correction of minutes of meetings, etc. With the development of the linking of computer with computer through the telephone network, electronic mail has also come into use as a personal form of communication which can be also used to provide access to stored information. Such facilities are not exclusive to scientific work, but are altering the information scene on which the next generation of scientists will have to act their parts. Publication is taking on a new aspect with the introduction of the Internet, the influence of which cannot yet be calculated.

However, remarkable though these changes may be, what has not speeded up is the money supply, to organisers and to individuals. This means that, as the changes we have described were coming about, the scientists were living in a world which expected them to conduct their professional business as if they were in a competitive world of commerce or industry or whatever. The pace of professional, individual scientific work was still determined by that mixture of luck, guile, and inspiration which has always been part of the 'scientific method'. That fine work of long-lasting value could be done with exemplary economy is shown by the International Biological Programme. How much more might have been done with more generous support we can only speculate after looking at its achievements.

Chapter 14

World projects and the environment

International Biological Programme (1964–74)

Seeing how widespread was the influence of the International Geophysical Year (IGY) it would be easy to get the impression that this was to be a general pattern of international progress. Not so. Inspiring and productive though it was, its achievements were linked with the nature of the sciences which were involved, and the materials of their study: mainly physical and geophysical. The biologists could emulate but not closely imitate.

The work and progress of IGY were closely watched by the biologists, many of whom held office in ICSU. Just as IGY began with people talking informally, so general conversation on social occasions led to the working out of a proposal for a biological analogue to IGY. Sir Rudolph Peters,[1] President of ICSU, chaired a meeting of the ICSU Officers in March 1959, in Cambridge, at Gonville and Caius College (where, for many years, F J M Stratton had run ICSU from his college rooms). On a train journey returning to London, Peters chatted with Lloyd Berkner who had been deeply involved in IGY, and Guiseppe Montalenti, President of IUBS, about the possibility of a biological exercise similar to IGY.

Peters then took this up with Ronald Fraser at the ICSU office (in The Hague at that time) and went on to Naples to develop the idea with Montalenti. They began to put substance into the idea with specific proposals about subject-matter, being particularly enthusiastic about the need to study isolated human populations before they were absorbed, culturally and genetically, into the general world population. Although this did figure in the final programme, the scope was eventually much broader. An Executive Board meeting of ICSU adopted the general idea of a biological programme, albeit ill-defined to start with, and planning was set in hand.

As with all such planning there were successive stages of topic-listing; first in 1961 a simple three-subject proposal (human heredity, plant genetics and breeding, modification of natural biological communities). Then began a stage (very usual in such discussion) of national emphasis on particular aspects. The Russians, for example, wanted to emphasise 'The biological basis of Man's welfare'. The initial enthusiasm was thus diluted at first by too many personal and factional concerns for concentration on limited topics. The world of biological study is very different from the world of physical study, the chief difference being that biological phenomena can take a long time to reveal their character. The plan of action of IBY therefore had to embrace a far longer time-span for action than IGY, which had provided the model.

IGY had meant a year that had been stretched a little but still a 'Year'. It was obvious that what would do in physics and geophysics, where there was much comparing of simultaneous observations, would not do in biology, where one would often need to compare successive annual cycles over several years. So IBY became IBP: *International Biological Programme*, the duration to be determined by the objectives.

C H Waddington, who was President of IUBS at the time, records how he felt at one time that the whole scheme was so lacking in focus that it ought to be suppressed.[2] However, defective though it was, it had gone too far to be abandoned, even though no one had yet given any sound thought to finance and organisation. So Waddington took the lead he felt to be his responsibility, and in January 1962 circulated a document: 'Notes on the selection of topics for an International Biological Programme'. Even though topics were still to be discussed, a title was settling into formal use. Waddington proposed that the pros and cons of the three topics being considered should be weighed up:

(1) Human genetics; a difficult subject to cover comprehensively because although factors like blood groups are easy enough to deal with the really important factors such as racial differences, or intellectual ability, are politically sensitive and improperly understood.

(2) Human population growth; this again seemed unsuitable for IUBS to tackle on its own. The basic facts were already well known, the detail was under intensive study by bodies with far greater resources than IUBS (e.g. the Ford Foundation). Moreover, the implications were politically sensitive.

(3) Man and ecology: the ways in which Man may modify the natural environment so that it can produce what he needs with the highest efficiency on a long-term basis.[3]

A planning committee met in Morges, Switzerland, in May 1962 hosted by Jean Baer, President of the International Union for the Conservation of Nature (IUCN).[4] Waddington was conscious of a lack of balance in attitudes: he himself was anxious to press the merits of a production ecological approach; many of those around him were more concerned with pure conservation. He had some sympathy from a representative of FAO, who was an expert in fisheries and therefore saw the need to consider productivity as a factor in the human relation with the rest of nature. The idea of ecology as the study of an energy balance had to be sold to some influential groups of biologists in the United States and in the United Kingdom to whom it was so far of minor interest. Eventually the Committee produced a programme which was an elaborated derivative of Waddington's preferred themes.

The organisation of IBP was to fall into 7 sections: 3 on biological productivity in terrestrial communities, 1 on fresh water, a fifth on productivity in marine communities, a sixth on human adaptability. A seventh was to deal with public relations and training, but these responsibilities were eventually taken over by each section itself. Instead a seventh section was established for use and management of resources which fell outside the provinces of FAO and WHO.

It looked fine on paper, but the biological world was far from uniform in its enthusiasms. In the United Kingdom there was as good deal of support. In the United States there was coolness in some quarters and even a degree of open hostility. However, some American ecologists did manage to get the ear of their colleagues, to persuade them to support the IBP proposals and to recognise that IBP was 'lifting a minor subject to a position of major status'.[5] In the long run it seems that the US contribution to IBP work probably exceeded that of any other country.

In early 1963, progress had been made to the point at which it was desirable to appoint someone to run the programme and the administration. The choice fell on E Barton Worthington. This able man had unrivalled experience of the right kind in Africa and in the UK, both as regards the organisation of scientific studies and as regards the management of large-scale enterprises.[6]

The years of planning from 1961 to 1964 needed to be followed by a three-year period referred to as 'Phase I–Preparations', to be followed by five years of 'Phase II–Operations'. Eventually the digestion of results were seen to need another two years of 'Phase III–Synthesis and Transfer'. This added up to a decade (which coincided with the International Hydrological Decade).

This was all operational. IBP needed lines of communication with the world community, the very purpose for which ICSU existed. Although not all the discussion had taken place under an ICSU umbrella, it was appropriate that IBP should be seen to have the status of an ICSU body, since ICSU was the only world organisation with the appropriate image. So ICSU set up a Special Committee to take overall responsibility for IBP. It was composed of representatives of those ICSU Unions which could be expected to play a part in the programme, and a number of individual scientists connected with countries and organisations each of which could be expected to play an active part somewhere in the programme. The Committee elected a President, Jean Baer, who was followed after five years in 1969 by François Bourlière. Worthington's position was defined as that of Scientific Director. He became a leading figure in more of IBP's functions than the purely scientific.

Differences from IGY, over and above the scientific, were apparent from the beginning. IGY had looked like a joint effort all along in spite of some hesitations. IBP suffered for a time from the initial caution of the Russians, and the downright opposition of some parts of the US community, an opposition which eventually faded and was replaced by a very large positive effort.

Third World countries might have been expected to be enthusiastic towards a project from which they might gain great benefits. This was in fact the case, but what could they do without internal funding? Goodwill does not pay bills. However, much IBP work was carried out in countries which could not afford to contribute much to the cost, which was borne mainly by the advanced countries. What is regrettable is that so much work in the Third World countries was done by expatriates, because in many of them there was as yet insufficient nationally developed scientific strength. This is one of the problems that will still not go away. Twenty years after IBP, the co-ordinated effort to overcome this problem got a name: *Capacity Building*.

IBP never enjoyed financial support as large or as widespread as IGY. No central fund was ever established, in spite of continued efforts. Costs are difficult to estimate, but Keay suggests that during the main operational phase something more than $40,000,000 per annum was expended by national bodies on research. Most of this came from contributions by member nations to research by their own scientists. The cost of administration of the international co-ordinating committee averaged no more than $200,000 per annum, a remarkably low figure.[7]

Although very different, it is interesting to compare some features of IGY and IBP. IGY had been thought about (as a Third IPY) ever since the Second IPY had been rounded off. It took seven years from the time of van Allen's dinner for Sydney Chapman in 1950 to the official opening date of observation. It was five years from the 1959 dinner in Cambridge to the formal launching of IBP, in its first General Assembly, at UNESCO in Paris, IGY and IBP differed not only in their duration, but in the recording of results. IGY was able to set up a system of reports in a standardised form issued by one publisher. The results of IBP were spread over a large part of the biological science literature. Much of the data resulting from IGY studies could be accommodated in the newly developing WDCs. Biological findings do not lend themselves to the compression one can achieve with physical data, and in many cases they require the preservation of objects and specimens. A comprehensive collection of original papers and reports was eventually housed with the Linnaean Society in London. Twenty-five numbers of a journal (*IBP News*) were issued. More than 200 international meetings were held. So much matter was produced that for it be at all accessible it had to be reduced to a detailed analysis and synthesis which were combined in a series of volumes.[8] There were also very substantial national publications, notably in the USA and the USSR.

One change brought about by the existence of IBP was a better balance in the policy-making of ICSU. The strongest influences in the General Assembly and in the Executive Board had long been that of the physical scientists. There had been discussions of biological matters since the foundation of IAA, but most of the prominent names and prominent issues for half a century of ICSU development had reflected this physical-sciences predominance. But IGY had played a part in promoting what one might call 'whole-world thinking'. As the support for IBP came more and more to the attention of ICSU enthusiasts, they were increasingly ready to think in terms of one world of science: the biological and the physical together. Moreover, IBP played a major role in turning world attention of other than scientists to environmental problems

It also emerged that the much desired co-operation between 'East' and 'West' was being encouraged by the recognition on both sides of the importance of co-operation in biological matters of great human interest.

The sixties were turbulent times. Arab and Jew fought each other and so found it difficult to preserve detachment at the symposium table. Central European scientific communities were disrupted by the pressures of the political forces in the USSR. IBP and ICSU did their best, often successfully, to overcome the consequent difficulties in getting a balance of joint action. However, they were at the same time making evident the practical significance to human welfare of IBP's findings. It became apparent that an intergovernmental equivalent to the non-governmental IBP was needed.

Here again we have an example of something substantial arising out of informal conversation, this time between Worthington and Michel Batisse (a leading natural resources scientist at UNESCO). As early as 1966 they talked around the possibility of a successor to IBP, possibly an intergovernmental one. SC-IBP did not think well of this idea at first, but it took root in the growing atmosphere of concern at the way human population might well outgrow resources.

Compared with the end-product of IBP, that of IGY was fairly tidy. IBP did not produce anything as clearly defined as the Antarctic Treaty, but what it did accomplish was to bring ecology forward as a primary concern of the biologist and, in addition, of the politician and the economist. While IBP cannot take all the credit, it was influential in bringing to the forefront of political and economic consideration the relations between man and the biosphere: On 4–13 September 1968 UNESCO held an Intergovernmental Conference of Experts on the Scientific Basis for Rational use of the Biosphere, under the chairmanship of François Bourlière. Not long afterwards plans were outlined for the intergovernmental Man and the Biosphere Programme which was formally launched in 1971. This can be considered the successor to IBP. The influence of IBP could still be felt 20 years later, in Rio de Janeiro.

Formation and development of SCOPE[9]

The development of SCOPE was in some ways typical of the growth of international scientific bodies. Like so much else in the growth of co-operative efforts, we can see the hand of one man: in this case T F Malone (USA) who drafted a proposal on environmental issues for submission to two of the Unions, IUGG and IUBS. These proposals were passed on to the Twelfth General Assembly of ICSU in Paris in 1968. A group was formed out of members of IGU, IUGG, IUBS, IUPAC, and SC-IBP. This *ad hoc* Committee recommended to ICSU that it form a Scientific Committee on Problems of the Environment (SCOPE), and lay out a structure and objectives. This envisaged, among other things, a global monitoring network, modest research operations, forward planning for an eventual world databank on environmental programmes, and a training programme for future environmental managers (with particular regard for the less developed countries).

In October 1969, the ICSU Executive Committee, meeting at Erevan in Armenia, approved the establishment of a Special Committee (that is to say a committee of limited life), a measure which was ratified by the General Assembly in Madrid in September 1970. An appeal over the signature of the President of ICSU (at this time V A Ambartsumian) was sent to academies and other national bodies, asking them to support the new SCOPE by designating a national liaison committee, appoint representatives, arrange to pay national dues, consider whether a voluntary grant in support were possible. From now on SCOPE was referred to as a Scientific (i.e. permanent) Committee.

A generous response came from the Royal Society of London, particularly through the energy and understanding of its Executive Secretary, D C Martin, a man who, over a career of thirty years (1947–76), contributed outstandingly to the work of ICSU, especially through IGY, COSPAR, IUPAC, and the SCFCS. For SCOPE, the Royal Society provided office space and the part-time services of a member of staff as Executive Secretary. R W J Keay, the Deputy Executive Secretary of the Royal Society acted as Treasurer and gave overall management supervision. Keay was able to do this on the basis of years of experience as a biologist working in developing countries. Any such organisation needs what is often called a Secretary-General, someone who will give consistency and momentum to the whole operation, and for this SCOPE appointed T F Malone. This wise and energetic man, who has been at the centre of activity in environmental matters for so much of his life, was able to call on the support of the Holcomb Research Institute of Indianapolis.

At the time SCOPE was being set up, the UN Stockholm Conference (projected for 1972) on the Human Environment was also being organised, and there were, naturally, consultations. These led to, among other things, a UN contract for SCOPE to undertake a review of environmental monitoring. This was accompanied by other schemes (such as one for the analysis of pollutants) so numerous that SCOPE had seemed in danger of becoming too diffuse and dilute for it to survive. There was also concern at UNESCO that some of SCOPE's programmes might needlessly duplicate parts of the MAB programme.

A good gardener deals with wasteful growth by judicious pruning; at its second Assembly at Kiel in 1973 SCOPE did just that: a new compact programme was agreed. It has been said that the new structure was drawn up overnight, a suggestion that anyone who has attended such a meeting will find easy to believe. The programme was slimmed down to six general topics:

1 Biogeochemical cycles.
2 Human impact on renewable natural resources.
3 Environmental aspects of human settlement.
4 Ecotoxicology.
5 Simulation modelling of environmental systems.
6 Communication of environmental information and societal assessment and response.

As time went on, some were broken down into more workable subsections, some objectives were achieved and rounded off, some became long-term ongoing projects, some petered out. There were many consequential programmes, some highly specialised, of universal significance, such as a review of the environmental consequences of nuclear war. In these SCOPE drew freely on relevant social scientists regardless of whether their associations were members of ICSU.

A short list of achievements will illustrate the outcome of this programme. The importance of the carbon dioxide content of the atmosphere is now a topic of common conversation: SCOPE launched its own comprehensive assessment of biogeochemical cycles, including likelihood of a carbon dioxide increase, at a time when these topics had not yet been explored in any significant way. In 1975 it began to promote studies of a number of the main elements involved in the biogeochemical cycles, taking account of the influence of human activities on them as well as their purely physical relations. It was by no means alone in this: the United Nations had founded its United Nations Environment Programme (UNEP) in 1972. SCOPE was able to contribute during the preparations for the Stockholm Conference (1972), especially in the area of monitoring, and later to UNEP's work, especially in the area of biogeochemical cycles in which extensive collaboration developed, including substantial financial support from UNEP.

These are large-scale matters. SCOPE has also concerned itself with matters which the individual human being meets under close and personal conditions. In 1976, SCOPE launched an ecotoxicology (a term invented by R Truhaut, for a long time the leader of the project) and health programme, which has been implemented in part through collaboration with the World Health Organisation, ILO, and UNEP, using their joint International Programme on Chemical Safety. This developed alongside the creation of an International Register of Potentially Toxic Chemicals (IRPTC).[10] The reports produced are aimed at industry, and at national and international bodies which have opportunity and responsibility to safeguard the health of employees and non-employed citizens. All forms of life are vulnerable to toxic hazards. The non-human must be protected for its own sake or to protect other creatures, which may be part of its ecological system or enter into a food chain.

Some of this must seem remote to people who have had no substantial scientific education because it must be discussed in technical terms. What everyone, scientifically educated or not, has worried about some time or another is the threat of nuclear war. SCOPE created a Committee for the Assessment of the Environmental Consequences of Nuclear War (ENUWAR) in 1983 to assess as comprehensively as possible the consequences of such conflict in terms of atmospheric and climate effects, ecosystem response, and the global food chain. It is believed that the wide media coverage of the findings often described in the press as 'nuclear winter'[11] may have been influential in negotiation on arms reduction. The resolution adopted finally by the UN on that subject was consistent with SCOPE's findings.

The ENUWAR project clearly shows the difference between ICSU's objectives and those of advocacy groups of scientists, of which the Pugwash Movement is a prominent example. This Movement was founded in 1957 as a follow-up to the Russell–Einstein Manifesto against nuclear war; as a body of distinguished scientists acting in their personal capacities it addressed governments on subjects like biological warfare, strategic arms limitation, and misuse of civilian nuclear technology. While recognising the importance of such actions, ICSU clearly chose not to mix presentation of scientific facts with advocacy statements; only with such a reserve could it maintain maximum credibility. The fact that one and the same scientist could be active in ICSU as well as in the Pugwash Movement made no difference.

Closely related is ICSU's policy, voiced for example by P Morel, Director of WCRP on the occasion of his retirement: ICSU considers that it is 'improper for specialists to make authoritative pronouncements outside their field of expertise'.

Discussion in, for example, the Pontifical Academy and in the committees of ICSU had established that such a study was needed and timely, and that SCOPE was the best body to carry it out. In setting it up, again T F Malone, though no longer an officer of SCOPE, played a major role, using his well-known persistent telephone technique, calling at all hours of day and night. SCOPE was soon convinced and put up $10,000 out of its reserve as starting-up money. Sir Frederick Warner, Treasurer of SCOPE, became the leader of the ENUWAR project, which, thanks to his experience in science and in industry, developed swiftly and effectively. In 3 years, 17 workshops and extensive correspondence took place in which 300 scientists from more than 30 countries were involved. The latter included all major nuclear powers and, significantly, Japan.

The results showed *inter alia* that a nuclear detonation taking place in the northern hemisphere was likely to lead to serious consequences such as famine in both the northern and the southern hemispheres. This mobilised the interest of countries in the southern hemisphere that hitherto had assumed they would remain quite free from results of nuclear exchanges among the northern superpowers. The scientific report published in 1985 was accompanied by a popular paperback version entitled 'Planet Earth in Jeopardy' (1986) which aimed at the public at large and had a wide distribution.

This politically most sensitive project showed also that ICSU was indeed capable of doing what it always professed it could do, that is, analysing problems objectively and dispassionately and presenting the results as they were, leaving it to governments to act on the scientific basis presented to them. Time and again when Warner was asked at press conferences what he felt governments should do his answer was: 'They should read the report', and when journalists said 'OK, but what next?' he always replied 'They should read it again.'

While we are concerned here with hard science, with the actual achievements of scientific workers or groups, it is important to see how SCOPE developed a philosophy of co-operation. We can all accept that work in science is different from work in other ways (e.g. art, literature, manufactures). It is useful also to think of work in environmental science as distinct in style from other science. SCOPE identified criteria for deciding what to work on.

It decided that it should not engage in bench or field work but in assessment and stimulation. It should engage only in studies which are agreed to be of major significance in world environmental problems, and are capable of resolution in a reasonable time with the funds, the people, and the management resources available. Since nations can handle national problems and Unions can handle single discipline problems, SCOPE should stress activities which are international, nongovernmental, and interdisciplinary. SCOPE should not poach on other organisations' preserves: there are already programmes for MAB, IUCN (International Union for the Conservation of Nature), and committees dealing with the Antarctic, with the oceans, and with space.

Especially after the Rio Conference there was an increasing need for independent assessment mechanisms for which SCOPE would qualify

Of all the constraints the one likely to be decisive in many cases was the availability of the right people.

Funding

What we can now say about SCOPE funding might well be applied, in spirit if not in every detail, to many other ICSU bodies.

Given the action criteria listed above, SCOPE was able to seek funds for many projects, great and small, and to arrange publication. Financial support came from many sources: the United Nations Environment Programme, UNESCO, WHO, the European Economic Commission, some of the great charitable foundations (Carnegie, Ford, Mellon, Rockefeller), and some businesses (Exxon, Mobil, Shell). In his 1987 article, G F White (then President of SCOPE) made a point that ought to be kept in mind in any consideration of the funding of international activities of the kind we are concerned with: the money equivalent of the time and voluntary services put into these projects far outweighs the cash contributions of the benefactors, however generous.

It seems to have been this last fact that persuaded many benefactors to give freely and without hampering conditions. They were also aware that the basic overheads were covered by national dues and, in the beginning, the Royal Society of London, and then by the Ministère de l' Environnement of France.

All treasurers throughout the ICSU family shared one post-war problem: that of working with colleagues in countries having non-convertible currencies. All manner of ingenious accommodations were adopted to bring the scientists from the main blocks together, in one or other territory. The United Nations provided some non-convertible currency for travel by nationals of non-convertible currency countries. Also effective was the hosting of workshops inside them. The involvement of scientists released in this way from currency restrictions was not only of practical benefit, but did much to maintain the international spirit in which the work was conceived.

In his 1987 article, G F White said:

> Serving without compensation, travelling on economy fares, living in modest quarters, and enjoying a minimum *per diem*, it is the stimulation and fellowship of congenial research work that brings them together. To be sure, an evening at the Bolshoi usually goes with a Moscow workshop and a clambake on the beach flavours a Woods Hole gathering: but the principal benefits which the participants take away are of new knowledge and fresh perspective on an absorbing problem of the time.

The numbers involved were considerable. White estimated that during the first 16 years of SCOPE's existence more than 1,400 scientists shared their experiences in that fashion. One cannot give an overall ICSU figure, but, since SCOPE is only one of very many ICSU bodies, the total number to whom these words might be applied is obviously very great.

The disregard of nationality and the insistence on freedom of movement of scientists are parts of ICSU policy. They guided SCOPE just as much. SCOPE also suffered from common difficulties, one of the most important of which was that of doing justice to science in developing countries, where the scientists were few in number and, being men of wide ability, were likely to be drawn away from science into administrative or governmental posts.

There is another way in which SCOPE is characteristic of ICSU bodies. It has kept its overall policies but has changed its ways of reviewing the way they are pursued; that is to say it has adapted its central administrative structure in the light of experience. An over-large committee became a General Assembly that delegated management to a small Executive Committee. From permanent commitments it had moved, by the time White wrote his appreciation of the position, to the execution of specific tasks with advisory committees. White wrote of the loss of stability which could follow specific-task style management, but since that time SCOPE has devised a form of supervision which aims at the best of both styles. It speaks now of *clusters*: areas of concern, made up of a number of components, which may well overlap to some extent, to their mutual benefit.

These clusters are Biogeochemical Cycles, Global Changes, Ecosystems and Biodiversity, Health and Ecotoxicology, all connected by the cluster Sustainability. Each cluster has a designated member of the Executive Committee as co-ordinator of several projects carried out in it, so as to promote coherence.

Influence of SCOPE

SCOPE contributed during the 1980s and early 1990s in more ways than one to the genesis of the major Earth System Research Programmes carried out by ICSU alone or in partnership with one or more UN organisations. As far as scientific substance is concerned, its work on the biogeochemical cycles, especially the carbon cycle, had helped turn the greenhouse effect from a scientific matter into an important item on the world's political agenda through the meeting at Villach (Austria) in 1985, financed by a SCOPE/UNEP contract; the results were published in the SCOPE Report series.[12] The meeting led eventually to the setting up by ICSU, WMO, and UNEP, in 1988, of the Intergovernmental Panel on Climate Change (IPCC) which periodically assesses climate change and its effects on behalf of governments, and is now responsible only to the member governments. Its chairman was B Bolin of Sweden, who was a major participant in SCOPE's work on the carbon cycle ever since it began in 1975.

Since SCOPE had been set up to assess existing knowledge and to draw up research agenda, it was not designed to follow up its own recommendations and do the required research. For this an altogether different body would be required, a fact which was appreciated especially in the USA. There, the increasingly harmful effects of humankind's activities on its own habitat caused widespread concern in scientific circles. At the same time, the need to exploit the enormous potential for satellite observation suggested that the time was ripe to put this to work for the benefit of planet Earth, that is to say, to add an additional motivation to the ongoing space programmes. This state of affairs helped, of course, to pave the way for the plan to set up the IGBP in 1986 at the ICSU General Assembly in Berne. Understandably, but initially to the dismay of the SCOPE Executive Committee, the first Executive Director of IGBP appointed was T Rosswall, who was then Secretary-General of SCOPE. He had to resign from his SCOPE position to set up the IGBP office at the Royal Swedish Academy in Stockholm which had generously offered to host the IGBP secretariat. The President of SCOPE (J W M la Rivière) was asked to chair the nominations committee for the (ICSU) Special Committee for the IGBP, which stressed the co-operation that would be required between the two complementary bodies. This co-operation gradually materialised from then on, SCOPE undertaking, at the request of IGBP (the International Geosphere–Biosphere Programme, 1986), to help design the research plans of some of the components of IGAC (International Global Atmospheric Chemistry Project), GCTE (Global Change and Terrestrial Ecosystems Project), etc.

Again looking at this from the point of view of the UN, SCOPE had some influence in originating an International Register of Potentially Toxic Chemicals (IRPTC). One of the first projects within SCOPE around 1970 was the idea of making a registry of existing and newly produced toxic chemical compounds, which would serve science by providing clues as to the relation between structure and toxicity, and provide governments with useful data on which to base legislation for storage, transport, handling, and use. This idea, first launched by C Levinthal (USA) was aired in 1970 and 1971 to the Preparatory Committee for the Stockholm Conference (1972), but it did not then find support and thus did not become part of the proposals laid down in the official conference documents. At the Stockholm Conference itself, the Netherlands Minister for the Environment put up a proposal to set up such a registry and won the day by a small majority. IRPTC became a successful undertaking under UNEP in Geneva.

This chapter began early in the 1960s, and has reached a point at which environmental studies can embrace whole biological systems or, indeed, the whole globe. The intense intellectual interest, and at the same time the daunting challenge of the present-day situation can be summed up in diversity. So important is this study in itself that in 1991 IUBS, SCOPE, and UNESCO (as MAB) baptised their joint programme DIVERSITAS. This programme put down its first roots 30 years previously in IBP. More recently it derives also from general concern about rapid loss of genetic resources.

We shall return in chapter 18 to a discussion of biological diversity which was a central scientific theme of the Rio Conference of 1992, where diversity of economic and political interests also played a confusing part. Before we reach that we ought to look at growing diversity within ICSU itself.

Chapter 15

ICSU and UNESCO

Relations with UNESCO

It is easy to write about 'ICSU in a vacuum' as it were: to pay little attention to the wide stage on which ICSU has acted. Indeed, reading the reports of General Assemblies could well support that attitude, since such meetings must necessarily turn inwards to a great extent on to the management of internal affairs and to reviews of what has been done since the last meeting. Were agreed tasks carried out? Has money allocated been well spent? This is what one might call 'club style' management, and very stable it can be for an exclusive inward-looking group which is interested only in permanence rather than change. However, ICSU is not at all like this, and a more careful study of the many annexes to General Assembly reports reveals three things going on at the same time: existing ICSU bodies, including ICSU itself, adapting to external circumstances; new ICSU bodies being brought into existence in response to changing needs; and new relationships being set up. In 1966 the General Assembly received 39 reports from associated internal or external bodies. The General Assembly of 1993 received 75. Some reporting bodies had disappeared, some were new.

Although ICSU has never been dependent on the existence of any other international body (League of Nations, United Nations, UNESCO, or other), and, as we have seen, originated and had established its major features before either UN or UNESCO was created, these two are the most important external international structures affecting the life of ICSU. Without UNESCO and the other UN bodies, ICSU would be very different, although it would still exist and be very effective.

The UN has created 17 other intergovernmental organisations[1] and acts with others having various degrees of autonomy, such as, in the financial field, the General Agreement on Tariffs and Trade (GATT) and the World Bank. The ordinary newspaper reader may not always immediately associate them with the UN, in the way he does when he reads of some action of the Security Council. Some of these UN organisations may play a part in ICSU's life, as for example in the effect of the World Bank on the life of developing countries seeking scientific advancement, or the effect on industrial development projects of the functioning of GATT. More obvious are links with Intergovernmental Organizations having clear scientific components, like the

World Meteorological Organization (WMO), the Food and Agriculture Organization (FAO), and the United Nations Environment Programme (UNEP).

One needs to take UNESCO first because, although ICSU has established important links with other UN organisations, it was with UNESCO that it first took centre stage in the post-war cultural scene. UNESCO has always been invited to send a representative to each ICSU General Assembly, and conversely ICSU has generally been represented at the General Conference of UNESCO.

The birth and early years of UNESCO are highly significant to the status of ICSU. Their subsequent relationship has not been that of dependency of one on the other either way, but rather one of mutual reinforcement and complementarity. For carrying out its mandate in science by itself, UNESCO would have needed a large number of first-rate scientists in most branches of science as staff members, which obviously would not be possible for two reasons: few good scientists care to work as administrators for prolonged periods, and secondly there was not enough money. Thus from the beginning ICSU was seen as an important advisory body to the Science Sector of UNESCO, which became (after Education) the second largest sector and, according to many, the most effective.

Furthermore, ICSU provided a most useful mechanism for interactions with the scientific community and for directing part of UNESCO's science budget to worthwhile projects, fitting UNESCO's mandate. While, after the Second World War, UNESCO's financial assistance had been crucial in setting up the first ICSU professional secretariat, future annual UNESCO subventions were spent in mutual consultation on projects of Unions and Committees. Although ICSU did not charge any overheads for the work, including the reporting entailed, the fact that ICSU had project money available for its members and committees helped to increase coherence and to widen the scope of initiatives and activities. This was especially important in the early post-war years; afterwards, funds from other sources increased, making the consistently substantial UNESCO subvention less and less important proportionately in the ICSU budget.

In addition, UNESCO can be looked upon as providing an intellectual and political background to the work of ICSU, and sometimes as providing a link with the strongly political world of the UN and its member states.

United 'Nations': what nations? The membership of the UN has changed radically since its inauguration in 1945. It began with 50 members. That number has increased more than threefold, the most important reason being the increase in the absolute number of states in existence, following the evolutionary break-up of the old empires, and the success of independence movements. Thus many new nations participate in decision-making in the UN, including decisions based on scientific study and discovery. This has increased the number of member states participating in the General Conference of UNESCO which has the last word. It has also provided ICSU with welcome opportunities for contacts with scientists in the Third World and also for taking note of their problems.

An attempt was made in 1952 at the Seventh Session of the UNESCO General Conference to lay the foundations of an extensive UNESCO conspectus of science through a UNESCO International Advisory Committee on Scientific Research. Its aim was to promote international co-operation between national research councils and centres of scientific research in fields of common interest. The Committee was set up in December 1953, with the President of ICSU as one of several ex-officio members. Others were representatives of the Council for International Organisations of Medical Science (CIOMS, founded 1949) and the Union of International Technical Associations (UATI, founded 1951). In 1962 it was agreed that some Unions whose programmes appeared directly related to items on UNESCO's own programmes in the natural sciences should also be members.

However, joint reviews of the state of scientific progress were considered essential and these were carried out by joint meetings of senior UNESCO science staff and officers and staff of ICSU. This constituted a strong personal link from 1946 onwards, but it was made even stronger in 1964 when the Executive Board of UNESCO 'welcomed the proposal that ICSU be invited to act as a permanent advisory body to UNESCO in the Natural Sciences Programme and noted the complementary proposal that the International Advisory Committee for Research in the Natural Science Programme of UNESCO be terminated'.[2]

One valuable product of the International Advisory Committee during its short life was the formation in 1956 of FAGS, one of the several systems by which ICSU sustained the collection and dissemination of data.

The first ten years of collaboration were not always harmonious. There was no real dispute, but rather a good deal of irritation felt by ICSU at UNESCO's failure to make the most of a golden opportunity for effective collaboration. This irritation was expressed by a man who, although experienced in articulating criticism, was equally experienced in suggesting and implementing ways of removing its causes. Harold Thompson, President of ICSU from 1963 to 1966, put the problem thus at the 1966 General Assembly in Bombay:

> At the Vienna Assembly it was agreed that ICSU should accept the invitation of UNESCO to become its principal scientific advisor, and this relationship is now quoted in many UNESCO publications. While I realise, of course, that scientists within ICSU are advising UNESCO on special matters, and some of the ICSU Programmes are receiving special financial help, which we much appreciate, I am uneasy about the position and do not feel that our attempt to coordinate activities and to ensure a planned distribution of funds is satisfactory, and I hope that in the discussions between Unesco and the officers of ICSU our advice may be found more acceptable.

Thompson spoke as if just for himself, but he was in fact expressing an opinion generally felt in ICSU circles. Things got better. At the next General Assembly in Paris in 1968, the UNESCO Assistant Director-General for Science (Professor A Matveyev) praised the friendly co-operation between ICSU and UNESCO and offered similes such as the two being sides of one coin or the two sides of a Möbius ring. ICSU could exist without UNESCO and UNESCO could exist without ICSU, but this is not the cause of any irritation.

Some joint action developed between UNESCO and individual Unions. In 1972 IUGS embarked on an International Geological Programme which had what one might now call the 'usual' components, namely, the promotion of geological research through internationally organised projects, and the advancement of geological science in developing countries. By 1993, IUGS could report that it had 55 active projects under way and that, although the subvention from UNESCO was small, it exerted a 'considerable leverage'. This good phrase says a good deal about those joint projects which have appeared to work well, not only in geology but in other subjects as well.

The value of the ICSU–UNESCO relationship comes out best in the way ICSU has so often produced an idea for a useful study, which, although it could develop it on its own, is seen to be of interest and value to UNESCO. In such cases it has often been possible to build a new structure for carrying out a new programme with UNESCO managing those aspects which have an intergovernmental bearing, and ICSU or an ICSU member managing the non-governmental aspects.

A case in point is that of the International Biosciences Networks[3] which were inaugurated as a joint ICSU/UNESCO programme in 1979, to help developing countries build up their own resources in biological science, research, and education. Regional networks have grown out of this, so one sees the international effort strengthening national facilities. There are now Asian, Arab, and Latin-American networks. In 1993, IBN merged with COSTED; UNESCO co-sponsors both.

Besides joint projects, each organisation carried out many scientific projects under its own flag. UNESCO was instrumental in setting up the European Centre for Nuclear Research (CERN, 1954); it developed the Man and the Biosphere Programme (MAB) and the International Hydrological Programme (IHP). It also strongly supported for many years the International Centre for Theoretical Physics (ICTP) at Trieste, a model institute for the education of physicists from developing countries, directed by the Nobel Laureate Abdus Salam (Pakistan) later President of TWAS, with which ICSU has a strong co-operative relationship. This is put here in terms of international co-operation, but it should always be borne in mind that no project like ICTP can take root or flourish without being welcome and receiving some essential degree of support of the host country.

UNESCO also promoted neurological research by supporting IBRO, now an ICSU Union, and cell research through providing funds for the International Cell Research Organization (ICRO), now an ICSU Associate. ICRO managed to organise 266 training courses in the period 1961–87 in 67 countries, most of them in the Third World. One outgrowth of ICRO/UNESCO work in applied microbiology was the setting up of a training and research network of the Microbial Resource Centres (MIRCENS), in the early phases strongly supported by UNEP, which is now collaborating closely with both IUMS and IBN/COSTED.

In all of these ventures, contacts with members of the ICSU family occurred, and similarly many pure ICSU projects contained elements of co-operation with UNESCO at the working level, SCOPE and IGBP being examples.

United Nations

To return to the UN, many cultures have an image similar to the English 'the moon and sixpence' to describe the way some small nearby thing may hide a larger more distant thing. We have mentioned UNESCO many times, but its importance in the scientific field with which ICSU is concerned may conceal from us the larger body, the United Nations, of which UNESCO is an agency. It may also conceal the cluster of other agencies which have been formed by it or are subsidiary to it. These also play some part in ICSU's history.

While UN bodies are sometimes in competition and feel compelled to defend their own interests, ICSU's role in joint activities with UN organisations is one of the neutral outsider who has no axe to grind. When, in some scientific issue, UN bodies are in competition, ICSU is often in the best position to make independent proposals which thereby become acceptable to all. Furthermore, when differences of opinion have arisen, the ICSU Secretariat has proved to be a neutral and congenial venue for a meeting to resolve them, which is taken advantage of from time to time. (In one case, a working lunch included two heads of UN agencies, and in the end, since ICSU has no butler, one of them helped serve food and wine to the assembled guests.)

The 17 'specialized agencies', to use the term of the UN Charter,[4] report annually to the UN Economic and Social Council. Some are older (in origin if not in present form) than the UN itself, for example the International Telecommunications Union (ITU), the Universal Postal Union (UPU), WHO, ILO. Others are new creations like the IAEA.

UNESCO is thus not the only United Nations organisation with which ICSU has built up good working relations, although it is the principal channel through which ICSU receives some international funding. It is going too far to say that the changes in ICSU's interest reflect world history, but it is worth hinting at. An early example of ICSU's work with a body with a strong scientific relevance is to be found in the problems of climatic change.

The concerns of WMO are closely related to certain areas of applied science. Here we have seen a kind of evolutionary process: an ICSU committee (the Committee on Atmospheric Sciences), which met in Geneva in 1966, developed some ideas which became the basis of an agreement between ICSU and WMO to initiate joint research on a world frame of reference. This was the Global Atmospheric Research Programme (GARP) which is discussed earlier in Chapter 12. A Joint Organising Committee (JOC) for GARP enunciated two objectives:

(i) to enlarge the physical understanding of the transient large-scale atmospheric phenomena so as to increase the accuracy of forecasting the weather – over periods from one day to several weeks – a global problem;

(ii) to enlarge our understanding of the structure and variability of the general circulation of the atmosphere so as to improve our understanding of the physical basis of climate.[5]

GARP marks an extension of ICSU's view of its responsibility: from the *scientists* of the world to the *science* of the world.

The origins of this kind of study lie far back, before the origins of ICSU itself. The geophysicists had long had an interest in the behaviour of the atmosphere, and IUGG played an important part in the second IPY and in IGY. There was already a guarantee of interest in the atmosphere as a separate study even before the launching of IGY, so ICSU was able to engage with WMO in the GARP project. There was launched what has been described as the largest experiment ever hitherto carried out: the First GARP Global Experiment or Global Weather Experiment (FGGE) which ran from 1 December 1978 to 30 November 1979. This may be taken as the starting-point for the gradual improvement in weather forecasting (including the achievement of successful five-day forecasting) which, quite apart from its scientific interest, has made a great difference to the lives of ordinary people who have never heard of the organisations which have had so much effect.

In 1979, ICSU and WMO signed an agreement for a joint World Climate Research Programme (WCRP) which formed part of a World Climate Programme (WCP). WCRP became operational in 1980 with the objective of attempting to find out to what extent climate can be predicted and the extent of human influence on climate. This kind of concern was to become outstandingly important in the following decade.

Another way of keeping a balanced view of the ICSU/UNESCO relationship is to look at a body that might claim to be older than the UN, that is to say SCOR (Scientific Committee on Oceanic Research), itself an example of that evolutionary process in the growth of ICSU. Its origins go back to the days when political movements were creating the League of Nations, just after the First World War. An International Association for Oceanography was formed in 1919 and continues today as the International Association for Physical Sciences of the Ocean (IAPSO), one of the Associations of IUGG. The idea of a biological study of the North Sea and the North Atlantic was not proceeded with but was taken over by ICES (the International Council for the Exploration of the Sea) which had existed since 1901 and had a detailed familiarity with these areas.

Neither of these quite matched the ICSU concern for a comprehensive view of research possibilities. IGY made a difference here, as it did for so many studies. Although giving much attention to physical matters, it did relate some of its studies of ocean properties and movements to biological consequences. By 1957 the need for a new organisation was clear, so ICSU created a Special (i.e. limited life) Committee on Oceanic Research. Its success led to its becoming a Scientific (i.e. permanent) Committee.

One of the major decisions of SCOR was very much concerned with human problems. It was agreed to embark on a programme of work in the Indian Ocean. Perhaps one can put too much meaning into this decision, but it is important in more than one respect. Most of the pre-war proposals had to do with the European–American interests. Now the interest moved to an ocean the littoral of which was of great importance as a food supply for nations which had previously played little part on the scientific scene. Moreover, it provided opportunities for nations which were far from the forefront of science to send their own nationals to sea in their own ships for scientific purposes. In 1960 UNESCO came into the picture, agreeing to co-sponsor the research, so in the same year there was founded an Intergovernmental Oceanographic Commission with which SCOR worked from then on.

USA and UK withdrawal from UNESCO

How pleasant it would be if one only had to describe aspirations achieved and opportunities seized. Unfortunately, there is more to UNESCO than the high-mindedness which gave it birth and is still its main driving force. It is an organisation of nations run by people with as much ambition, prejudice, conflict of loyalties as any other large organisation. In 1985 criticism of UNESCO led powerful forces in the USA to persuade the US government to withdraw (which it did from the beginning of 1985). A year later the UK and Singapore followed suit. This did not mean withdrawal from ICSU. Both the USA and the UK provided ICSU with extra grants to make up for the decrease in the UNESCO subvention caused by their withdrawal.

Politics is not just a succession of gestures. The USA and the UK may have withdrawn from UNESCO, but they still need some of the things it stands for. A few lines of communication are kept open, one of the most important being ICSU.

This short discussion has moved irregularly between ICSU and UNESCO. It has had to attribute initiatives to each of them in turn. Over the period from 1946 to 1980 a question seems gradually to emerge: is ICSU just a body for the support and encouragement of dispassionate research, or are there also other imperatives at work like the need to bridge the science gap with the Third World, or the need to ensure the survival of humankind? Like UNESCO in the hands of its member states, ICSU's course of action is determined by the General Assembly of its Unions and national members. To see if the question has a meaning we need to look at the internal changes in ICSU: its components, its structure, and its self-examination.

Chapter 16
Membership

Definitions of membership

Many of us know the experience of finding that a few friends share some of one's own ideas and enjoy discussing them in informal gatherings, these conversations developing later with regular schedules and recognised conventions. It is also a common experience that such a successful association may find it necessary to turn conventions into rules and statutes. Later still one finds that these need redefinition and clarification from time to time. So it was with ICSU. The first statutes, elegantly expressed though they were, fell short of unambiguous definition in many ways, especially in respect of who or what might be a member.

The early statutes did not attempt to define 'Union'. The 1931 version referred to 'the various international Unions', as if they had some necessary permanent existence. The key words in the titles of the early Unions were familiar in common speech (except perhaps for URSI). The statutes seem to have taken it for granted that any body seeking admission would already be of long standing. They did not define 'nation' or 'country', but did make it clear that Dominions and 'Protectorates' (a League of Nations term) would be independently eligible so long as it could be shown that the applicant supported independent scientific activity. The statutes have become more precise.

There has been a subtle change in the wording for 'national membership'. The 1931 statues said:

> (III) The International Council of Scientific Unions consists of a national scientific organisation from each country which has adhered to the Council and of the International Unions.

And further:

> (IV) A country may join the International Council either through its principal Academy, or through its national Research Council, or through some other national institution or association of institutions, or, in the absence of these, through its Government.

The 1952 Statutes had a few but insignificant changes, but by 1994 there was a new definition of some weight.

(III.7) A Scientific Union Member shall be an international non-governmental professional organization devoted to the promotion of activities in a particular area of science for at least six years,

A footnote defines *international*:

In these Statutes and Rules of Procedure, international bodies are taken to mean those bodies to which appropriate organizations in all countries of the world are eligible to adhere.

Further by 1994:

(III.8) A National Scientific Member shall be a scientific academy, research council, scientific institution or association of such scientific institutions. Institutions effectively representing the range of scientific activities in a definite territory may be accepted as National Scientific Members provided they can be listed under a name that will avoid any misunderstanding about the territory represented.

A National Scientific Member had also to have been in effective existence for at least four years.

The collection of the individual histories of all the members of ICSU, Unions and other, would amount to a large encyclopaedia, and, in any case, be too fragmentary to give a general impression. No two Unions are the same in structure or constitution, nor does membership mean the same thing to all of them alike. However, the gradual addition of new Unions to ICSU membership is a rough and ready index of the development of science at large. The seven founder Unions, around 1925, covered the limited traditional range of sciences which were systematically studied wherever there were universities and government research institutions. They were not organised in the same way; each had its own way of dealing with the subdivisions of its subject. None was comprehensive or covered all the topics potentially appropriate to its subject scope.

Similarly, national membership has never meant quite the same thing to each national member. In the broadest terms, the early national membership was limited to the advanced countries and their larger dependencies. The founder Unions were limited to the traditional physical and biological sciences, the major advances in which had been made in the advanced countries. After 1945, national membership followed the political development of a world of which one could speak of as divided into three, but which was moving towards better intercommunication under the United Nations than had ever been possible under the League of Nations.

At the General Assembly in Sofia in 1990, some uncertainties had emerged in the discussion about the weight each group of members should have in decision-making. Perhaps the impending drastic increase in national membership was causing anxiety that the 'balance of power' would shift too much towards the national members. One delegate put it that they were after all, less concerned with science than the Unions. This raised a storm of protest. The Standing Committee on Structure and Statutes was asked to look into this, and their pertinent proposals were accepted by the General Assembly in Santiago in 1993. The text reads:

> Article 6: . . . Members may adhere to ICSU in one of two categories:
> (a) Scientific Union Members
> (b) National Scientific Members

indicating that neither of them had a monopoly of being scientific.

> Article 24a: In order to ensure equality of votes of the two categories of membership in the General Assembly, each National Scientific Member has one vote, and each Scientific Union Member has that number of votes which is equal to the number of National Scientific Members, divided by the number of Scientific Union Members, except in votes concerning finance, in which case each member has one vote.

Voting by proxy or by mail was to ensure that absence from a meeting would not upset the new system.

Changing character of Unions

The General Assembly of 1949 had spent a long time considering a distinction between General and Specialised Unions and recognising this in the voting powers allotted to each. Bit by bit the blurring of the boundaries of Union specialisms created a need for means of bringing Unions together in special projects. 'General' and 'Specialised' had diminishing relevance, and eventually the voting distinction disappeared. New kinds of *ad-hoc* bodies were being created for the purpose of examining new problems. ICSU was then, as now, always available to offer or adapt means of unification, and to create new ICSU machinery.

Those who might want to make a new Union had several patterns to follow, but the Union type of structure did not meet all needs. In 1956 the first of a new kind of ICSU activity was conceived. It became evident that a service was wanted for the purpose of unifying the results of observations made in a number of fields, the relation between which was quite as important as the separate fields themselves. The pattern had already been shown, to a limited extent, in the Second International Polar Year of 1932. Each field was becoming so extensive that piecemeal recording and dissemination might be uneconomic or even doomed to be self-defeating.

Union membership spread into fields connected with medicine. Links with industry, which had been implied in the titles of IUTAM (International Union of Theoretical and Applied Mechanics, IUPAP (International Union of Pure and Applied Physics) and IUPAC (International Union of Pure and Applied Chemistry) also developed. This expansion is still, however, more to be seen in special commissions within existing Unions than in overall Union activities, and is held by many to be inadequate. (This point is considered in respect of the Ringberg meeting of 1985.)

The people who gave, and give, their time to the Unions were, and still are, as temperamental as any others, scientific or not. They have always identified with their work and with their peers, so it is not unexpected that there should be differences of community sense as between practitioners of one science and another. If there were not, there would never need to be a vote in an ICSU General Assembly. These differences of personal identification are significant in considering the place of science in the world: it is easy for the 'man in the street' to think, or the journalist to write, of all 'scientists' as being the same in outlook and temperament. They are not. The man who wants to measure the shape of the earth is a different kind of man from the one who wants to measure the shape of a fish. Put this in the plural and you begin to describe different communities of scientists. They can be strangers to each other, and it was one of the values of ICSU (and remains so) that it was able to create the common ground needed to turn strangers into friends and colleagues.

We cannot describe the origin and development of every one of the Unions but a few will show how different these have been. We can take one aspect only in each case, remembering that the story of development of each Union is complex.

IUPAP

In 1922 a number of physicists attending the Second General Assembly of IRC in Brussels decided to form an international Union for Physics. Thirteen nations immediately decided to take part, and a steering committee was formed. The first President was William Bragg (whose son, Lawrence Bragg was to form the IUCr 25 years later).

The First General Assembly of IUPAP 1923 concerned itself with, among other things, the need to accompany each scientific paper with an abstract by the author, and with means for disseminating these abstracts widely to the abstracting journals of each country. (Among the participants who were concerned that this be done on account of the unreadable explosion of scientific papers was Marie Curie, who must have been particularly aware of the expansion of a subject from small beginnings to a wide-ranging study.) At the same General Assembly an appeal was made for help in reconstituting Japanese library collections lost in the recent earthquake.

In 1931 the General Assembly heard of the transformation of IRC into ICSU and decided to adhere to the new body. The international congresses of physics continued up to the 1939 war, considering for the most part the new physics that was constantly on stream.

The French H Abraham, Secretary-General from 1922 to 1943, died in a concentration camp. Paul Ewald acted as Secretary-General from 1946 to 1947, at the same time as he was involved in the development of new organisations for X-ray crystallography. Some were not happy at the creation of separate Unions which had a physics basis. Ewald urged vigorous action in putting IUPAP on a business-like footing, with proper archival records and a permanent secretariat. This was also the time to take account of the new relationship between ICSU and UNESCO, and to begin to establish the Union's own links with UNESCO. The number of interunion Commissions was increased

In 1951 it was urged that the Union increase its activity in more specialised areas of physics and put less emphasis on the social effects of science. This was different from the attitude many members of ICSU and its Unions were to adopt later on. The 1954 IUPAP General Assembly did not like an ICSU proposal to put Unions into two different groups, Special and General, a proposal which did not last long.

The Tenth IUPAP General Assembly saw a recurrent problem. The Republican China operating from Peking was admitted as a member, but withdrew on finding that the Taiwan China had also been admitted. The Chinese problem lingered in many parts of ICSU for long after this. IUPAP was one of those Unions who often showed their own interest in the freedom of the scientist. In 1984, for example, it went on record as supporting appeals for the freeing of Sakharov.

New Commissions were formed bringing hitherto speciality subjects like solid state physics to the fore. In keeping with a general ICSU trend, a commission on teaching was formed. One small item from the 1963 IUPAP General Assembly serves to show the pervasive nature of Union work. It heard that 41 International Congresses in its field had been held since the last General Assembly. The same sort of thing was reported at the General Assemblies of other Unions as well. ICSU was not the initiator of all this work, but it was the centre for the distillation of its consequences.

IUBS

The development of IUBS has been well described by its Executive Director (T Younes), so we shall follow his account.[1]

Biologists have always been internationally minded, personally as travellers in search of material far from home, and conceptually in their efforts to systematize and name the objects of their study. It is altogether appropriate that the records of the largest biological study ever carried out, the IBP, are deposited in the Library of a distinguished Society named after Linnaeus.[2] Biologists were thus among the first to see the advantages of international congresses, in Botany beginning in 1864 and in Zoology in 1889. Although botanists and zoologists lived in worlds apart for a long time, the general theories of biology which started with and followed Darwin led not only to an interdisciplinary approach to observation, but also to new theoretical systems with genetics and cytology very prominent.

An International Union of Biological Sciences was proposed at the First General Assembly of IRC in 1919, and was formally constituted in 1923 and admitted to Union membership of ICSU in 1925. Its work, until 1939, was not extensive: mainly limited to botany and to co-operative bibliographic work. It cannot be said to have encouraged new thinking in biology, and like other Unions it had to lie low from 1939 to 1947.

After the war it changed radically, under the influence of officers who were leaders in such non-traditional disciplines as genetics, physiology, and embryology. IUBS took a prominent part, creating the International Biological Programme, which was taken over by ICSU, but IUBS can take much of the credit for initiating the modern approach to ecology.

Younes' account is very frank about the efforts made to keep a balance between traditional and non-traditional areas of biology, which failed because the new disciplines and the challenge of new applications of biology in agriculture environmental issues and the needs of the Third World were seizing majority interests. The projects which were generated were too big for IUBS itself and were, for the most part, taken over by ICSU and absorbed in SCOPE or IBN.

To put it in colloquial terms, IUBS decided 'If you can't beat them, join them'. In 1979 it modified its constitution to enable it to develop its own multi-disciplinary projects, each one with objectives, which, although limited, were such as to attract the co-operation of a wide variety of biological effort. There was a *Decade of the Tropics,* a *Bio-Indicators* programme, and programmes in *Biodiversity* (which is concerned with the study of existing diversity) and *Biological Complexity* (which is concerned with changes in global systems). This culminated in the setting up of the main research programme DIVERSITAS in collaboration with SCOPE and the MAB of UNESCO.

With this move into new scientific programmes came a substantial increase in available funding. A permanent secretariat had been set up, a very necessary requirement for the management of any organisation like IUBS. One change had been the separation and independence of the International Union of Biochemistry in 1955. (It changed its name in the light of the extension of its interests in 1991, becoming the International Union of Biochemistry and Molecular Biology.) One clarification has been the reassessment of the importance of taxonomy as a support for other types of work, rather than as an introvert commitment of traditionalists.

During these changes IUBS has accumulated a network of constituent commissions and committees and a membership of over 90 other bodies having international standing of their own.[3] The kind of evolutionary change it has undergone is typical of many Unions. We can pick out of these other Unions one that had been part of IUBS but in course of time became an ICSU member in its own right, namely IUMS (International Union of Microbiological Societies).

IUMS

Its very early experiences are typical of many Union developments. A historian of IUMS (Eric Kupferberg) begins his account by observing that, since the forerunner of the International Union of Microbiological Societies was founded over sixty years ago, it has undergone several name changes and eight major re-organisations. It has become a large, administratively elaborate organisation, but its growth did not proceed with ease. For a long time it lacked permanent headquarters, staff, funds, and statutory authority. National societies and federations of them existed but were uncoordinated. A few individuals, however, brought about great changes.

Up to 1912 the major outlet for microbiological reporting had been the Congresses of Hygiene and Demography. They were not revived after 1919. Microbiology expanded from the 'golden age' of Koch and Pasteur and their immediate successors. Good work was done in more and more countries as the relevance of microbiology in medical, industrial, agricultural, and social contexts was increasing. The number of microbiologists increased to the point at which an international body was obviously needed. It was founded in 1927 as the International Society for Microbiology (ISM), the leading proponent of its creation being the young Frenchman R Dujaric de la Rivière. He encouraged some influential seniors to back him up. Others got interested, and at the International Conference on Rabies in Paris (25–30 April 1927) an International Society for Microbiology was founded. Kupferberg claims the Rabies Conference as the first reconciliation of French and German scientists since the First World War. If so, perhaps it signifies that nothing is more productive of alliance than facing a common enemy! (Unhappily the harmony did not last long under political changes in Germany.) Under the Presidency of the Belgian Nobel Laureate, Jules Bordet, the First International Congress of Microbiology was mounted in July 1930.

Bordet sought two things of importance. One was that the statutes of any body which might eventually be formed should have a declared intention of fostering peace among nations. (Events, as we know, showed this to be a hopeless aim.) The other was that, by getting a number of the most pre-eminent men in the main fields of microbiology to be associated with it, the body should be seen to have great authority. With Dujaric de la Rivière as General Secretary he succeeded in both.

This First Congress agreed to establish the International Society for Microbiology on a permanent basis. If an organisation like the ISM is to be credible, it must show continuity. This was established by the formation of a Commission, to work between congresses. The subject it addressed was of great importance: bacterial nomenclature, a subject which was in disarray, not being adequately dealt with by any of the existing nomenclature schedules.

It is not always apparent to outsiders (even working members of the profession concerned) that setting up an international congress is an expensive, elaborate, exhausting, even heart-breaking business. This is true of all the organisations discussed in these pages, so we can take the ISM experiences of 1933–5 as no more than characteristic. An invitation to meet in Germany in 1933 was accepted, the President to be the German bacteriologist, Martin Hahn. The world economic situation imposed a postponement to 1934. By then Hitler was in power. Hahn had a very little Semitic blood in his veins. He was forced out of his German office and left Germany, to die not long after.

The Congress was to be postponed yet again to 1935, but the German National Committee withdrew its membership. This might have meant collapse, but a United Kingdom invitation saved the day and the Second International Congress for Microbiology opened on 25 July 1936. It was clear that microbiology had changed, both by its own elaboration and by the appearance of links with other sciences, physical as well as biological. It was seen that in such a congress what mattered most was not the papers but the personal contacts. (Anyone who has served on a grant-giving committee knows how hard it often is to persuade the non-scientific members of this.)

The Congress took what appears now to be a retrograde step. It changed the name of ISM to IAM, International Association of Microbiologists, and abandoned the idea of a permanent headquarters. It was to be a confederation of national organisations, without individual membership, and with no provision for activity outside of or between congresses. We cannot tell what effect this might have had. The Third

Congress was held in the United States, opening on 2 September 1939. War broke out, the UK declaration being made on 3 September. The state of tension had prevented many delegates from being able to attend. Others who had got there had to hurry home. The limited Congress survived, with many contributions pointing to future progress, but the invitation to hold the Fourth Congress in Denmark turned out to be a sad mockery. The Fourth Congress was in fact held in Copenhagen, but not until 1947. At a Congress in Rome in 1953 the title was again changed to IAMS (International Association of Microbiological Societies, and policies changed in the direction of more activity between Congresses (which were themselves getting bigger and bigger, almost unmanageably so). IAMS had been a Section of IUBS and in 1967 changed its status to that of Division, an increase in status and dignity. But microbiology grew so strongly in the post-war period, not least because of the visible success of antibiotics, that IAMS developed ambitions to become an independent Union rather than a part of IUBS. This it achieved in 1980, and after prolonged discussion it became a member of ICSU in 1982 as IUMS.

IUNS

The stresses of war made people think in new ways about familiar things, including nutrition. It is interesting to see the emergence of a new union taking place in parallel with a growth in public awareness of the kind of issue which lay behind an area of scientific progress. There had always been some popular appreciation of the need to adopt an attitude to food and well-being, as shown in a music-hall song 'A little of what you fancy does you good'. Gradually the various factors in nutrition had been identified and their interrelations discerned. The basic energy need had been first measured in the 19th century. With the development of the chemist's skills in analysis, the knowledge of the composition of the many substances participating in the chemical aspects of animal and vegetable lives had promoted the recognition of the importance of the emerging science of biochemistry. Nutrition, however, remained on the borderline between the scientific and the popular. Efforts to put it on the same basis as other economically significant studies failed to get official recognition until the anxieties of war gave the study a new public face. When this had happened it was possible for the community of scientists working in this field to set up an international body, a Union, which would prove to be acceptable as a member of ICSU in 1968.

IBRO

IBRO is one of the most recent admissions to Union membership of ICSU. The International Brain Research Organisation is an example of a body which had developed to a high degree of influence and world-wide recognition before becoming a Union member of ICSU (in 1993). It was also one based on recent scientific advance. The instrument on which their work relied, the electroencephalograph, had first been described by Hans Berger[5] in 1929. The origin of IBRO can be traced back to a meeting of encephalographers in London in 1947, not really very long after the opening of the field. This meeting led to the establishment of an International Federation of Societies for Encephalography and Clinical Neurophysiology. In 1960 IBRO was created; in the first phase it had considerable support from UNESCO.

Then its scope expanded into the much wider area of general neurology, until it was decided in 1982 to modify its constitution to recognise this change. IBRO became an Associate Member of ICSU in 1976, that is to say, communicating with ICSU, taking part in some of its deliberations, but not having yet any authority to influence its administrative or financial decisions. Over a period IBRO's own style changed. Up to 1982 IBRO had existed to provide links between numbers of people who got to know each other through academic contacts. This individual membership was quite small compared with the large number of neuroscientists throughout the world. There existed, however, many other neurological societies, national and international, which could manifestly benefit from some sort of co-ordination. IBRO therefore changed its constitution to allow other organisations into membership. Each individual member of each of these member bodies could then, if he or she desired, became a member of IBRO. This constitutional change was not all that easy to bring about, since IBRO is established by Act of Parliament in Canada, and the change was slow and expensive.

IBRO has its own ways of communicating with the Third World, and went through a period of criticism for what some thought was an excessive concentration on it. This was brought into balance with the new style of constitution. IBRO also showed originality in creating a category of Supporting Members, who could be pharmaceutical manufacturers or scientific instrument firms, showing more originality in this than ICSU was able to do in its own search for 'partners'.

The body that ICSU could see in IBRO changed therefore from a modest learned society to a very large professional community, which had a far wider scope than only interest in the expertise of the EEG. By 1993 IBRO had developed sufficiently strong consciousness of its world standing to seek full membership of ICSU as a valuable support. It was admitted at the General Assembly in Santiago (October 1993)

Personal membership was important in the development of IBRO. It has been less so in most other Unions since their activities have mainly depended on specialized groups. However there are exceptions such as IAU which has had individual membership for a long time and IUPAC which introduced it later with success.

National scentific membership

We have kept referring to ICSU's 'national' membership but, as we have said, 'national' does not mean 'national government'. Who can speak, to the outside world and within a country, for its general body of scientists? The answer is different in each country and reflects its political constitution. There are the old academies (creations often of monarchs); academies of the 19th century, acknowledging the existence of a scientific community; academies of the 20th century acting as government research organisations; embryonic academies, creations of the past few years in developing countries. ICSU has also had to consider whether status could be granted to some organisations such as a university within a country where there was as yet no other significant body of scientists.

Nearly all of the advanced countries had become national members of ICSU soon after the political pattern had settled down after the Second World War. Germany (the BRD) came in 1952 with the Deutsche Forschungsgemeinschaft as its representative body, Germany (the DDR) not until 1961 with the German Academy of Sciences in Berlin as its representative. From the unification movement of 1990 onwards, the DFG became the sole representative. Irish culture has for centuries been held in high esteem, but it was not represented formally in ICSU until 1978, perhaps because a national constitution for an Irish state took a long time to reach a sufficiently settled condition.

The status of national scientific associate was useful in bringing into contact with the wider world many smaller nations, especially some in rapidly changing Africa, for example, Burkina Faso (the former Upper Volta). There could be problems like those in East Africa. The 1978 General Assembly heard how there was confusion over representation of Kenya, Uganda and Tanzania by a single Academy. Eventually Kenya was admitted as National Member, represented by the Kenya National Academy of Sciences. Uganda and Tanzania are for the time being members only of a few Unions.[6]

The most striking changes in membership have followed the transformation of the USSR. Science has flourished in most of the constituent republics of the USSR with some degree of independence. (One President of ICSU, V A Ambartsumian, and one Vice-President G Skryabin, were Georgians and proud of it.) However, the USSR insistence that there should be one, and only one, international representative scientific body had prevented scientific groups within republics from standing alone on the international stage. After the reconstitution of the republics each could make its own scientific character visible, and there were many additions to ICSU's roll.

By 1993 there were admitted as national members: Latvia, Lithuania, Armenia, Belarus, Georgia, Estonia, Moldova, Ukraine, and also Russia. Uzbekistan became a National Associate. The entry of so many bodies, not only into ICSU, but onto the world scientific stage, posed problems of communication and of support. The older national academies, for example, set up committees and appointed specialists to provide the necessary liaison. These new developments were looked at by optimists as a new resource, but there were others who preferred to consider the needs of these nations as imposing a moral and material burden on the long established members of ICSU.

New kinds of membership

So, repeatedly, there surfaces an issue which one would suppose had been resolved in the early days of ICSU and indeed of IRC: what is a national member? There were some who urged that a national member should always be a government or be determined by a government. Against this was the view adopted by the founders of ICSU: that a

national member of ICSU should be an academic association existing within a particular country and responsible to its scientific community. This was agreed to be the
correct interpretation of ICSU's status as a non-governmental organisation. It was this
insistence on a strict non-governmental criterion that was to help in the resolution of
the Chinese problem ten years later.

In 1984 there was added an Associate Member which added a new dignity to ICSU,
the Third World Academy of Sciences (TWAS).[7] This body originated in informal discussion at a meeting of the Pontifical Academy of Sciences in Rome in October 1981.
An influential group of scientists from India, Pakistan and South America prepared a
memorandum which led to the founding of the Academy in Trieste on 10–11 November
1983. Its official launching took place on 5 July 1985, by the then Secretary-General of
the United Nations, J Perez de Cuellar. The relevance of TWAS to the event was that
its birth confirmed that ICSU had to come to terms with a world which was changing
in personal terms as well as in physical and industrial terms. These Third World acad-
emicians were not only persons of consequence, but now a substantial group. If they
were to appear separate and unusual their existence as an academy would be divisive.
Fortunately TWAS has had the wisdom to state clearly that it will take care not to
duplicate in its programmes work of other organisations with common aims. This
means that the differences between TWAS and other scientific bodies that had originated in the European/Atlantic world can be seen as lying in the variety of science and
its applications, not in the variety of race or culture with which the concept of the Third
World has been so strongly associated. (The first President of TWAS was Abdus
Salam.)

It has yet to be seen whether TWAS was an isolated case or the beginning of an
important new trend. It was classified by ICSU as a Regional Associate, although its
'region' was world-wide. It was appropriate, because there was already another
'regional' association, a good deal older, the Pacific Science Association founded in
1920 as the Pan-Pacific Science Conference, and admitted to ICSU in 1970. No
member of TWAS (and as an academy it has only personal members) was born in a
rich country, although some resided in such countries later in life. In 1988, TWAS
helped in the establishment of TWNSO (Third World Network of Scientific
Organizations), and, in 1993, of TWOWS (Third World Organization of Women in
Science). In 1986 there were admitted as Regional Associates of ICSU the Academia
de Ciencas de America Latina, founded in 1982, and the Federation of Asian Scientific
Academies and Societies which had been established in 1984.

ICSU welcomed the development of international scientific communication and
mutual support through the admission to associate membership of these new groupings, but there were those who advised that a careful eye be kept on the balance of
power and influence within them. Any grouping of such entities can suffer from one or
other member exercising too great an influence. This goes along with the other political realities which ICSU has to face in maintaining its non-political stance.

But some groups cut across any political boundaries. For example an important
International Scientific Associate is the International Institute for Applied Systems

Analysis (IIASA) founded in 1972 and admitted in 1987. With TWAS it was one of the co-sponsors of ICSU's ASCEND 21, held in preparation for the UN Rio conference of 1991.

These are all examples of 'partnership', the idea that came much to the fore around 1985. This was the year of the Ringberg meeting, a meeting which attempted to look at ICSU as the whole it had become and the different whole some thought it should aim to be.

Chapter 17

Ringberg (1985) to Visegrad (1990): self-examination

ICSU Reflecting

In August 1983 the ICSU Executive Board broached the idea of a special meeting to examine a whole range of principles and prospects, strengths and weaknesses, so as to provide guidance for future development. The lead was given by Sir John Kendrew (President) and T F Malone (Treasurer). Two years later, on 7–9 October 1985, 45 participants met in the Schloss Ringberg near Munich, which had been placed at ICSU's disposal by the Max Planck Institute. Financial support had been provided by the Alexander von Humboldt and Volkswagen Foundations, and by the Deutsche Forschungsgemeinschaft (ICSU's national member for the BRD). As always, a lot of paper was circulated to members before the meeting, which was more than usually necessary because the delegates were not all ICSU aficionados, and it was important that all should be speaking to a common background. By deliberate intent there had been invited a number who were more familiar with other kinds of international scientific co-operation, and able to bring a critical light to bear on matters with which ICSU itself had become too habituated for its own good. The title chosen for the discussion was *International Science and the Role of ICSU: A Contemporary Agenda*.

One can make too much of the word *agenda*. It too often used to signify just a list of topics to be reported or aired, without commitment to action. Its older sense was what was implied at Ringberg and at the Visegrad meeting of 1990, namely, a list of things needing to be done.

There were several intentions at Ringberg. One was to look at a possible future orientation of ICSU and to consider how to make this orientation effective. Others were how to keep up a continual review, how to improve communication among members and with service bodies. As always in such affairs, questions of finance and of staff had to be considered. These initial thoughts were rather vague, but they crystallised into a good working programme with a spirit of realism, summed up in F Mayor's quotation[1] of a personal remark of Indira Gandhi 'I have already too many books and reports. Please do not produce another thick document; produce just one solution.'

There were four Working Group Topics for discussion:

    (a) the diverse roles of science;
    (b) the intellectual spectrum and reach of ICSU;
    (c) past and future global projects;
    (d) ICSU's external partners.

The general atmosphere was one of a dignified concern for science, with a realistic appreciation of what it had become during the 20th century. Several speakers pointed out that, while formal scientific communication had changed greatly, with a proliferation of the printed word, the most fertile interchange was still by word of mouth, cross-fertilisation, and recognition of other people's work often running well ahead of published reports. Lighter touches were rather rare in this serious atmosphere, but the Russian soil-scientist, V Kovda, achieved one by observing that the universal language of science appeared now to be 'broken English'.

The diverse roles of science

There was a general recognition that there was neither truth nor value in any distinction between *pure* and *applied* science, but that there were *science* and *the application of science*. This had several consequences for ICSU. There might be no need to distinguish sharply between 'pure' and 'applied' in describing the scope of a Union, but there was a need to insist that the interest of a developing country and its resources should not be confined to supposedly 'applied' science but should concentrate on basic science just as much as was done in the developed countries. When the basic science was taught, and made the centre of research, the applications would follow.

It was pointed out that there was a temptation to engage in projects which should really be left to intergovernmental organisations, such as the World Bank. Such projects might, for example, be efforts to affect the economic life of a country by the introduction of some new but known technique believed profitable or productive.

Besides this warning against wrongly conceived development programmes, there was the risk of an 'internal brain-drain': the benefits of higher salaries leading to scientists moving into administration and so depriving the country's scientific community of talents that should best continue to be applied to science. This talent should not only be applied to new research but also to the training of the country's next generation of scientists of its own creation.

Granted this caution, it had to be recognised that many problems needed the co-operation of behavioural and social scientists. This must not be sought at the end of some project, merely for comment, but integrally throughout any study which could be seen from the beginning to have social implications. Moreover these could and should usefully be studied in parallel with the 'hard' ' science.

Throughout the Ringberg meeting, challenges were often posed, one of the harshest being this: 'For whom has science been successful if two-thirds of humanity has no benefits from the knowledge gained by science?' Some reasons for this were offered: for example, that a scientist in a developing country (say, in Africa) often saw a career in the furthering of some field of science as needing to be pursued in the scientific tradition and culture in which that field originated, which is likely to be some developed country (say, in Europe). Such a scientist could not therefore establish the recognition and prestige that would follow successful pursuit of science originating in the conditions of his or her own country. He would be all the less able to show his fellow citizens that science was their common concern. It follows that those who are not familiar with science are inclined to see it as something alien to their national needs and ambitions.

Such a problem was not limited to the problems of developing countries. The scientists of the advanced countries also found public understanding eluding them. Questions of information and communication in their many forms constantly recurred at Ringberg. One quotation from the official report sums up the final view: 'ICSU should rethink the role of scientists in the process of communicating or explaining science to those outside science.'[2] Many reasons were offered, one of the most important being that the time between discovery and application gets shorter, so there is less and less time between something new being learned and the subsequent involvement of the general public in its consequences.

There will probably be no end to debates about the motivation of science. There are those who hold that scientific investigation is impelled merely by curiosity. Others hold that there is always an element of practical problem-solving in any research. And there are those who believe that both kinds of motivation exist side by side, sometimes in the same person. These debates are not by any means pointless, because in any particular set of circumstances the balance of opinion may affect programming, funding, and development. The debates have a bearing on how ICSU behaves.

The intellectual spectrum and reach of ICSU

In 1946 the report of the General Assembly of ICSU and the attached reports of the 7 Unions listed altogether about 100 fields of activity, each having a committee or formal directing group. By 1982 there were 20 member Unions having more than upwards of 400 committees, formal groups, or bodies closely associated with Unions. A great many of the groups had to do with topics which did not exist in 1946 (for example, space science, recombinant DNA, much of solid-state physics, laser technology, etc.). And this without counting the ICSU Special and Scientific Committees with their subgroups.

The scientific world in which ICSU lived was therefore in constant flux, with new fields emerging and new groups forming to specialise in them. This may lead to fragmentation, to unnecessary creation of new journals, to competition for resources. It was felt that ICSU could counterbalance these forces of fragmentation by developing new means of drawing together narrow disciplines, old as well as new, into new interdisciplinary projects or continuing committees.

Several factors had been making for change. One was the progressive penetration of technical devices into daily life (the car, radio, new materials, air-travel, etc.). Another was the great progress that had been made in countering infectious disease. Another was the expansion in the complexity of the operations of scientific research itself. Another was the growing consciousness of the ordinary investigator of the economics of his occupation. And, since many a scientist earns a good living in other ways than academic research, one had to look at the growing importance in the scientific community of the industrial scientist. Ringberg never came fully to grips with this last issue.

To the number of these interest groups one must add the very large number of bodies other than Unions with which ICSU had co-operative association, such as the Scientific Associates. Some of these were quite complex in structure, like the International Union of Food Science and Technology. Some represented sciences new to the generation of the Ringberg participants, like the International Federation of Societies for Electron Microscopy.

ICSU then, had done two things: it had taken on communication with the great variety of new organisations which had grown directly out of the original member-Unions. It had also taken on a considerable number of representatives of the sciences which had originated in new discovery in the previous 40 years. Some experts had developed their ICSU connection within existing Unions: some, like radiation protection, needed separate representation.

In many of these organisations the current work originated in 'applied' science. In 1985, speakers at the Ringberg meeting recommended that greater interest in applied science (which does not necessarily mean scientific work within industry) was desirable and necessary to broaden the scope of ICSU's work. In the event, progress has not taken the line of 'doing more of the same', of merely multiplying committees and commissions to look at applied science. It has come through a change of attitude, from the institutional to the identification of global problems. 'Application' has come to signify something more than the mere utilitarian or profit-making. The environmental seems to be taking first place instead.

The discussion of the variety of science within ICSU opened out to a wider view: should ICSU seek some link with the social sciences, perhaps through ISSC, the International Social Science Council.[3] This body had existed since 1952, an assemblage of some 14 international non-governmental organisations representing social science disciplines. The administration of these bodies was rather loose, and ISSC lived a somewhat saprophytic existence, relying almost entirely on receipts from UNESCO for services under contract. An autonomous organisation of national social science bodies from some 30 countries had been set up and had become a member of ISSC.

The language of the Ringberg report is double-edged: 'The state of methodology in some social sciences and the strong cultural and political influences that prevail in the application of the social sciences in different parts of the world demand special efforts for meaningful collaboration with the basic natural sciences.' Evidently Rosenblith and Thurau[4] were not persuaded that the social sciences as they saw them in action in some areas, could be treated in the same way as the basic physical sciences. Nevertheless they were certain that interactions between natural and social science disciplines needed to expand. They were careful to speak with greater respect of the Humanistic Sciences and of the International Council for Philosophy and Humanistic Studies (ICHPS) which had been founded in 1949 on a UNESCO initiative, and had main aims which harmonised with the cultural elements in the aims of ICSU. They did not go into detail, as if the interests of ICHPS were of ancient respectability as are those of the sciences.

The next field of common interests to be mentioned was engineering. Although there had been some common ground between the scientists trained for the most part academically in the basic sciences and those whose learning had brought them into the world of construction and manufacture, there was still little communion between the scientists of ICSU and the engineers. All the same there was some, and there were signs that there might be more. After all many ICSU scientists had engineering degrees or other qualifications. The engineers had developed two main international bodies, a World Federation of Engineering Organisations (WFEO), and UATI (Union of International Technical Associations).

The reporters of the Ringberg discussion put it briefly 'The tradition of international technical co-operation is much less strong in the engineering area primarily because of proprietary concerns.' This was too simple an explanation: the Ringberg disputants did not seem aware that the history of engineering has too many examples of the spread of ideas from individual to individual for commerce to be the whole story. It is more likely that the language and apparatus of communication in engineering inhibits long-range mutual criticism and analysis to a greater extent than in the basic sciences. As engineering became more and more a matter of formal analysis, and more and more open to interpretation in terms of the basic sciences, so communication became not only easier but more necessary.

Be that as it may, the engineers had not yet formed any body with the comprehensive embrace of ICSU. Within ICSU, IUPAC and IUB had established statutes which permitted association of profit-making companies. The scientific view of a profit-making concern is different from, but not alien to, that of the academic, and the scope of these two Unions recognised this. The reporters were far-sighted enough to see that the emergent biotechnology was going to raise new problems and opportunities in this area of pure and applied science.

Past and future global projects

To many present at Ringberg, the major issue confronting ICSU was the potential contribution that the world scientific community might make to the illumination and resolution of emerging global problems. The session on Past and Future Global Projects began with a reminder that benign global conditions favourable to human life would eventually be challenged by the proliferation of the species and the alteration of the environment. The predictions quoted were that the world population will reach 10 billion over the next 100 years with 1.4 billion living in the developed countries and 8.6 billion in the presently less developed countries.[5] The special place of ICSU, it was asserted, came from the fact that, although there are very many international organisations ('a vast array' was the expression used), ICSU is unique in its universality, diversity, non-political tradition, ability to enlist top intellectual resources, and a demonstrated capacity to manage, with great prudence, an annual budget of (in 1985) six million dollars. However, all these attributes needed careful critical re-examination to protect, or, if necessary, justify them. (It has to be said here that concentration on 'Global Change' was in fact to be a distinguishing feature of ICSU development in the following decade.)

The discussion of past successes and failures identified several topics for special study: focussing on specific goals, developing a multidisciplinary approach, integrating national and international efforts, strengthening national groups, planning for a timeframe which extends beyond the usual limits of political planning, (inevitably) finance, flexibility in incorporating into an active programme developments in science and technology which occur while the programme is running, utilising the most modern effective communications and data systems available at that time to aid the programme, standardisation of temporal and spatial scales so as to make results in different disciplines truly comparable, maintaining uniformity of method and approach over different regions or countries and as between different disciplines.

This reads as enthusiastic, but in fact there was caution: 'Should ICSU undertake Global Programmes? The answer appears to be a qualified "yes".' Problems to which ICSU could contribute usefully would need careful identification and strategic planning so that only those which arose because of manifest gaps in knowledge should be addressed. Even when this was done, the emphasis should be on interdisciplinary problems beyond the immediate province of existing organisations.[6]

What's in a name? The consideration of interdisciplinary action led to discussion of the title of ICSU. Was it just a matter of Unions? Would not a more descriptive title be *International Council of Science*, and would this be more commanding and authoritative? The question remained in the air: what mattered was not what ICSU was called, but what it accomplished. ICSU remained ICSU.

The discussion recognised that, whatever its title, ICSU was not self-contained: much of its activity was in conjunction with other bodies, IGOs and NGOs for the most part. The question arose whether there might now be other kinds of large and influential body with which ICSU should seek a *modus vivendi* and for each link a *modus operandi*. Some large Foundations sought to take a closer interest in the detail of the use of their contributed funds than hitherto. Many science-based industries acted on a world scale and had large numbers of scientists in their employ. The old respectful, respectable image of the scientist was out of date.

One serious problem was that of ICSU's relations with governments: should it, as an NGO, seek direct contact with a government which was active in a field with a visible scientific relevance? One answer to this was to improve the level of interaction with and amongst the national members of ICSU.

If the Ringberg meeting demonstrated anything, it was that to speak of the 'living machinery' of ICSU has a good deal of justification. Mere formality played little part in the discussions. The picture developed at the end was of an ICSU organisation that, while stable enough to survive in the form the participants saw before them, was never-theless confronted with the need for change, was capable of change and had the means of change, in its existing constitution.

The meeting could not foresee the political changes that were to come about in the next few years. Who could? They were to affect membership, participation, finance, and the free circulation of scientists. One might say that ICSU had always been pre-pared to come to terms with scientific discovery, but it was not organised to meet radical changes in public attitudes to the subject-matter of science. In 1984 it was fash-ionable to talk about mankind being murdered by nuclear war. By 1993 the threat had changed. It was no longer war, but suicide, death self-inflicted by destruction of the habitat. The global debate needs another chapter.

ICSU's external partners

The Ringberg discussion of ICSU's External Partners could have been sharp and stim-ulating if the idea of external partners had been an entirely novel departure. However, ICSU already had a wide range of external relations of which the ICSU participants in the discussion must have been well aware. The visitors needed to have this explained before they could join in a discussion which was really looking for some extension to a system of partnership that already existed. To understand this one needs to look at the situation then existing. Of some 5,000 international organisations listed in the *Yearbook of International Associations*[7] some 10% were members of the ICSU family, co-operating at different levels. Importantly the statutes also provided that, in order to carry out its tasks ICSU may:

> maintain relations with the United Nations and its agencies, and with other interna-tional intergovernmental or non-governmental organizations.

It may also:

> enter, through the intermediary of the national adhering organizations, into relations with the Governments of their respective countries in order to promote scientific research in these countries.

At the time of Ringberg, ICSU had formal arrangements with seven UN bodies and one regional intergovernmental organisation. It had informal arrangements with a number of other UN bodies, IGOs and NGOs. The Unions' committees and commis-sions also had a variety of arrangements of all degrees of formality.

The firmest arrangements were with the UN bodies, some formal, as laid down in those bodies' constitutions ('consultative status', 'official relations', etc.), some looser and of shorter life. Very close were those with WMO, with which it could efficiently co-sponsor large programmes of very specific content like IGY (1957–8), GARP (1967–80), and the successor to GARP, the World Climate Research Programme (1980–).

The closest relations, as might have been expected, were with UNESCO: almost daily, and detailed. At the time of Ringberg, this part of ICSU's life was clouded by the disturbing extraordinary political move by the United States we have previously mentioned.[8] Objections had been raised there to the way the Director-General, A M M'Bow, had been managing the internal administration of UNESCO and, it was claimed, had been giving its cultural activities an unacceptable political bias, had manipulated internal democratic decision-making, and (an accusation always, and sometimes justifiably, aimed at unpopular administrators) committed extravagance. In spite of strong argument from the US scientific community that such a move would be detrimental to US national and scientific interest, the US government went ahead and withdrew its membership and funding as from the beginning of 1985. The next year, the United Kingdom followed suit. (Neither nation has resumed membership at the time of writing). The loss of funding to UNESCO was very large, about a third of its previous expectations. The reduction in the UNESCO subvention to ICSU that resulted from this was made good by direct grants from both countries.

In spite of their withdrawal, both the USA and the UK could continue to support projects which originated with UNESCO but had a sufficient degree of independent status to be beneficial and to justify some funding. This strange situation was not resolved. There continued to be strong US and UK representation on ICSU and Union committees and associated bodies. The US and UK governments continued with their posture of protest without now having voices within the governing body of UNESCO. ICSU tried to formulate proposals encouraging the support of the Science Sector, but took care not to respond to hints that it might itself replace UNESCO as the principal international scientific agency.

The conclusions of the analysts of the discussion on external partnership were not very startling. There was even some doubt as to whether ICSU should or should not interact directly with governments. Leaving this aside it was agreed that ICSU should use its influence to persuade governments of the importance of basic/fundamental research. Raising the visibility of scientific discovery and of ICSU itself was considered important, but no concrete suggestions were made about how this might be done consistently on any scale.

'ICSU's interactions with and among its national members should be improved and it should consider methods for extending contacts with other scientific communities.'[9] This means, of course, better involvement of scientists of developing countries (which was already taking place) in the planning and execution of many programmes.

A suggestion with more apparent force in it was that ICSU should be prepared to carry out factual scientific studies on proposed scientific activities prior to their adoption by governments or intergovernmental bodies. Quite how ICSU was to appear sufficiently authoritative in such enterprises was not clear, but the thinking seems to have been mixed up with the idea, referred to above, that ICSU should become an International Council of Science, broader and sterner than the present limited body. (Over many debates in many ICSU assemblies and committees ideas of this kind keep coming up and keep being dropped.)

One of the participants was F Mayor, who was to become Director-General of UNESCO in 1987. He emphasised the importance of ICSU aiming always at promoting good science with the minimum of political interference. Partnership with governmental and intergovernmental bodies was essential, but it should be on the basis of communication, not domination. So Ringberg did not get very far with the idea of new forms of partnership.

What did come out of Ringberg? Rather little it seems at first sight. However, from then on in the debates in General Assemblies and in other ICSU meetings where policies were discussed, the main Ringberg issues came up again and again as guidelines. For example, the setting up of an Ethics Committee was discussed year after year. A new title, *Science International*, and a new style intended to increase ICSU's 'visibility', were found for the ICSU Newsletter. The recommendation to increase work with developing countries hardly needed making: it was so obviously desirable to build on existing work. The work in education for young people already in hand was intensified but the expression of the need to improve opportunities for women in science remained a mere expression of sentiment.

Visegrad: ISIP

One aim was pursued: it had been suggested at Ringberg that there be a follow-up meeting at which partnership should be a central theme. This took place at Visegrad in Hungary in May 1990, with the formal title *International Science and Its Partners* (ISIP). There was a larger attendance than at Ringberg, wider in its range. Notably there were some industrialists and several who, although nominally 'scientists', worked in an engineering milieu. The spread of ideas was therefore thought likely to be wider than at Ringberg and to lead to more constructive conclusions. In the end it produced no novel ideas, only encouragement to ICSU to pursue more energetically ideals which had in any case been expressed in many a General Assembly or other meeting.

The convenor of the Visegrad meeting, Professor W A Rosenblith, who had contributed a great deal to the Ringberg meeting, had worked hard and with vision in this follow-up, but his concluding observations cannot conceal an air of disappointment that no brave new ICSU world had been opened up. He deserved better for his labours.

Reading the official report of the Visegrad meeting as soon as it was published, one would have had little idea that the political world was in a state of convulsion around it. Although one speaker put it bluntly that great things were happening, the political comments were more of the nature of asides. The tension in Czechoslovakia (which has come to a head in separation into two republics since Visegrad) was mentioned as being felt already, but Germany and the Berlin Wall (which was already down, 10 November 1989) were not mentioned in the report.

The President of the Hungarian Academy of Sciences did refer to the changes within his own country and related them to the Gorbachov-led changes in the USSR. He stressed the way in which the internal affairs of one country are bound to affect those of neighbours, and related the condition of science within the USSR and Hungary to the general manner of government and economic management prevailing up to the time of change. He emphasised that changes are not to be taken for granted as being beneficial, but that time must be allowed to have its effect in producing the hoped-for results of closer contact between his country (and others like it) and the West. His speech was illustrated by the fact that during the Visegrad meeting, the democratically elected Hungarian Parliament met for the first time.

There were some engineers and industrialists present who had considerable experience in the international aspects of their callings. All looked to an improved interaction between ICSU and industry, but none had a practical pattern to offer. Robert M White (who became President of the National Academy of Engineering in the USA) played a prominent part in the debates. A recommendation to form an Industrial Advisory Board resulted, which carried ICSU a step further than Ringberg. (Despite extensive efforts by the Executive Board, no Advisory Board had materialised by 1994. Two obstacle were later identified by the Executive Board: the world's industries are only very loosely organised internationally and it is very difficult to find concrete scientific projects that require joint efforts of science and industry).

A E Pannenborg (Philips Company, The Netherlands) was worried that politicians wanted support for scientific work to be justified too often on the basis of seen economic benefit. His observation was brief, and deserved to be more prominent.

H Ausubel (Carnegie Commission on Science and Government (USA)) also offered a whole clutch of ideas about ICSU's world role, suggesting as alternatives pressing UNESCO to make its 'S' work better; taking the 'S' out of UNESCO altogether and creating a new world scientific body within the UN complex; starting a new intergovernmental organisation for science, based on the national science foundations; reconstituting ICSU on a bicameral basis, with a new relation between its national and scientific components.

This exploration of partnership began with the best of intentions, but it seems to have been weakened by its abstractness. At the very time it was being mounted, there were other influences at work to set up a different kind of survey, an examination of problems in which partnership would arise from perceived need rather than from desire for formal change. ICSU, like other NGOs had to set out, along with hundreds of other organisations, on the road to Rio.

Chapter 18
The road to Rio and beyond

Most people are aware of the existence of a problem of the environment, narrowly human or more broadly biological. Some realise that an important driving force for change had emerged in the sixties and had gained strength ever since: the recognition that the earth is finite and that humankind is endangering its own habitat. What is not so widely appreciated is the extent to which this had alerted scientists and policy-makers alike and led each into actions which run parallel. Even less appreciated is the very large numbers of scientists involved and the complexity of the organisations for study and action which have grown up in a quite short time. Let us look at what happened between 1972 in Stockholm and 1992 in Rio de Janeiro.

The scientists reinforced their efforts to address the questions posed by the Earth System: how it functions, how it is influenced by human action, to what extent it can be 'managed' by preventive and remedial action. Thus a new motivation for scientific endeavour added itself to the existing ones of satisfying curiosity and the desire to develop applications for industry, agriculture, medicine, and defence. A new, additional obligation came into sight: the responsibility to provide policy-makers with the essential scientific basis for developing and implementing strategies for safe-guarding the global life support system. The underlying premise was, of course, that man can repair a system that he has damaged provided he understands how it works and the damage is not inherently irreversible.

The policy-makers from their side realised that problems of the global environment, as diagnosed by science, were inserting themselves into the international political agenda, from world-wide pollution by chemicals as brought to popular notice by Rachel Carson (1963)[1] to the possibility of total climate change brought to their attention by the Villach Conference in 1985.[2]

Two streams of activity developed: the scientists started up programmes of assessment, diagnosis, and research with respect to the global environment, while the policy-makers began to promote that research and foster political will among the nations to take preventive and remedial action. This led to major UN conferences, new UN bodies, and, ultimately, to conventions. We have yet to see what will come of it all in the very long run, but we can see some first steps.

As international policy-makers had come to look more and more to the outcome of scientific research and observations as an indispensable factor in their considerations, the need arose for mechanisms that would connect international science and international policy-making in a concrete and practical manner.

Since ICSU had its dual roots in the world's scientific community (through Unions and academies) and was the sole international scientific body with substantial experience in the required type of research programmes, it became the obvious candidate to serve as a partner to the policy-makers with the tasks of creating new scientific understanding through research and observation and of communicating the results. Since these tasks were in harmony with ICSU's statutory mandate, ICSU accepted step by step the growing responsibility of wearing the new mantle that was gradually slipped onto its shoulders by the sequence of events.

One could argue that, had ICSU not existed, something like it would have had to be invented to fill the need. But one could also ask the question: was ICSU, as essentially a volunteer organisation, fully equipped to assume this new role, or in other words, should it have left this to another body. And finally, could not the UN itself have set up the machinery to provide for its scientific requirements?

While it is too early to answer all of these questions, it is evident that the UN does not have experience in carrying out on its own large and complex interdisciplinary research programmes, and has no structure specifically designed to do so. However, it can provide the intergovernmental element in programmes that need it and collaborate with the scientific community as a partner. This partnership formula has shown itself useful, as demonstrated in GARP, WCRP and more recently, the three Global Observing Systems in which ICSU is a partner of equal standing with one or more UN organisations. Its non-political nature and lack of bureaucracy help ICSU to fulfil its specific role of bringing objectivity, high scientific quality, and vitality to the programmes which thereby gain in effectiveness and credibility.

ICSU and the UN system responded, each in its own way, to the pressing world problems of the last decades, and for better or for worse joined, when useful, in partnerships without giving up their independence. Both evolved along separate tracks that ran parallel but were interlinked where it counted. One track is marked by UNCED, the other by IGBP.

The UN system: from Stockholm to Rio

It is important to recognise that the actions of the UN system, with regard to the global environment, were driven by the results of scientific endeavour in combination with such political realities as world population growth and the aspirations of all nations to reach or maintain a high standard of living. The Stockholm Conference on the Human Environment (1972) was a first step for placing the problems on the political agenda, and as a result many countries created a 'Ministry for the Environment', while the UN set up its United Nations Environment Programme (UNEP). This was only mandated to promote and co-ordinate the environmental work of the existing UN agencies. It was the first UN body located in the developing world (Nairobi) and, despite its small budget, it achieved some remarkable results like, for example, the Convention on International Trade in Endangered Species (1973), the Mediterranean Action Plan (1975) to improve the environment of the Mediterranean basin, and the Montreal Protocol (1987) for the protection of the ozone layer. UNEP at all times maintained close liaison with the scientific community, and participated as a partner in many projects of members of the ICSU family, the earlier mentioned work of SCOPE on monitoring and on the biogeochemical cycles area being cases in point.

During the 1980s it became clear, however, that the Stockholm Conference and the resulting actions of governments and the UN system would not be effective enough to turn the tide and realistically to address the globe's predicament. The Brundtland report 'Our Common Future' (1987), commissioned by the UN, stressed that the urgent needs for economic growth and development in the South could only be met within the limits of the earth's life-support system if world-wide collective action for wise resource use, leading to 'sustainable development', were taken. This led to the UN Conference on Environment and Development (UNCED, Rio de Janeiro, 1992) which aimed to place sustainable development not only on the political, but also, especially, on the economic agenda. It had pragmatic objectives: an action plan, *Agenda 21*, and several conventions. Based upon the findings of science, it was a political conference that attempted to make the governments act upon the insights and projections science could offer.

The organising Secretary-General was Maurice Strong (Canada). Strong had served in the UN, and in a number of other international bodies concerned with energy, finance, and environmental matters. He had acquired a good reputation as an organiser of the kind of conference envisaged. Particularly relevant was his experience in running the Stockholm Conference in 1972.

While the collaboration of ICSU with UNESCO continued, a new partnership had developed in the sixties with WMO in the area of climate research, starting with the Global Atmospheric Research Programme (GARP, 1967). In 1979 the first World Climate Conference created the World Climate Programme which included the WCRP as its research component and successor to GARP. In 1988, WMO and UNEP created the Intergovernmental Panel on Climate Change (IPCC), mandated to assess present scientific knowledge, possible impacts, and response actions. In 1990 the Second World Climate Conference (in which ICSU was a full co-sponsor along with WMO, UNEP, UNESCO, IOC, and FAO) gave a new impetus to climate research and observation and could be considered as part of the preparatory process for the Rio Conference. In contrast to the first one, the Second World Climate Conference was more than a scientific/technical meeting: it also comprised a meeting of heads of state and ministers of 137 countries, indicating the extent to which the concern of policy-makers for climate change had increased in one decade. This led eventually to the Framework Convention on Climate Change, signed in Rio and ratified in 1994.

ICSU: from GARP to IGBP

The above rough outline of developments in the UN system is useful background for understanding the parallel and connected developments in ICSU. Let us go back to 1967, to GARP, the origin of which was described in chapter 12.

In the sixties it became clear that new computers and new satellite observation technology brought a much better understanding of the dynamics of the atmosphere within reach that could lead to significant improvement in weather forecasting. The UN General Assembly passed a resolution (1966) urging ICSU and WMO to undertake the required research. The first planning meeting held in 1967 in Skepperholmen (Sweden) developed an outline of a programme which was to comprise the acquisition of a huge new data set that had to be both global and simultaneous. This required massive commitments from both governments and scientists. A Tropical Experiment was planned for making up for the lack of knowledge in the tropics; it was to be followed by the Global Experiment lasting about one year which would require new observing systems employing ships, satellites, aircraft, and buoys. The objectives of GARP were to study the transient behaviour of the atmosphere as a basis for improved *weather* forecasting and to study the general circulation of the atmosphere to improve understanding of the physical basis of *climate*.

By 1980 its success led to its being superseded by a new structure, but GARP had not only proved to be a model for co-operation of non-governmental (ICSU) with an intergovernmental (WMO) organisation; it also worked on the basis of a formula that was to be adopted later by WCRP and IGBP, in which a central scientific, planning, and steering group designed a programme framework – in iterative consultation with scientists of the participating countries – into which nations could deploy their costly equipment and manpower to maximal advantage.

While the most prominent result of GARP was the capacity for extended *weather*-forecasts, its objective of better understanding of *climate* gained in weight during the seventies through the concern about climate-induced disasters and increasing awareness of possible climate change through anthropogenic greenhouse gas emissions. This led in 1979 to the first World Climate Conference, essentially a scientific–technical meeting, which initiated the World Climate Programme (WCP) in 1980, comprising as its research component the World Climate Research Programme (WCRP) as successor to GARP but with new objectives: to determine the extent to which climate and climate variations can be predicted, and to assess the extent of man's influence on climate. In view of the role of the oceans as the 'slow part' of the climate system (in contrast to the 'rapid part', the atmosphere), the WCRP had to become more interdisciplinary so as to include oceanography, initially through ICSU's SCOR and later (1993) by adding the Intergovernmental Oceanographic Commission (IOC) of UNESCO as a cosponsor to WMO and ICSU. For a truly comprehensive treatment of climate, an even wider scientific view, even extending into chemistry and biology, was going to be needed. This was to be taken care of by the IGBP, which in this respect would complement WCRP and collaborate closely with it. The shift from *weather* to *climate* and towards more interdisciplinarity also required a shift in the WMO 'constituencies' in WMO's member countries, traditionally consisting mainly of the meteorology offices without participation of ministries of environment and/or scientific research.

From the projects undertaken, one study TOGA (Tropical Ocean and Global Atmospheric Study) achieved remarkable results with the inter annual climate variation linked with the El Niño[3] phenomenon; already farmer communities in, for example, Peru, find it to their advantage to select the crops they are going to plant on the basis of preliminary experimental TOGA predictions.

WOCE, started in 1990, promised to be the largest oceanographic programme ever undertaken, aiming to achieve one single consistent representation of world ocean dynamics, invaluable for improving climate models. GEWEX (Global Energy and Watercycle Experiment) studies the global hydrological cycle in relation to energy fluxes including the important role of clouds in the radiation balance of the earth. It will contribute to the understanding of the distribution of rainfall and the availability of moisture and usable water. GEWEX closely collaborates with the BAHC (Biospheric Aspects of the Hydrological Cycle) core project of IGBP. The Secretariat of the WCRP was located in the WMO Secretariat in Geneva. The Joint Planning Staff Director was P. Morel (France). Meanwhile the work of SCOPE in the seventies (see chapter 14) and early eighties reaffirmed the need to take the climate issue seriously. At the Villach WMO/UNEP/ICSU Conference on the Greenhouse Effect (1985), recommendations were made to governments including:

> the understanding of the greenhouse question is sufficiently developed that scientists and policy makers should begin an active collaboration to explore the effectiveness of alternative policies and adjustments. Efforts should be made to design methods necessary for such collaboration. Governments and funding agencies should increase research and focus efforts on crucial unsolved problems related to greenhouse gases and climate change.

There followed that dual response that was referred to in chapter 14: ICSU, WMO, and UNEP setting up the Advisory Group on Greenhouse Gases that was replaced by the Intergovernmental Panel on Climate Change (IPCC). During its work IPCC was to invite and receive considerable contributions from WCRP and IGBP. The second response was the Second World Climate Conference which reinforced the mandate of the WCP and the WCRP.

At the same time SCOPE's work had helped promote a broadening of climate change to the wider issue of what was later called Global Change. In 1979, (at the Fourth General Assembly of SCOPE in Stockholm, June 1979), M K Tolba, Executive Director of UNEP, and G F White, President of SCOPE, issued a declaration on Global Life Support Systems[4] in which they stated:

> The time is ripe to step up and expand current efforts to understand the great interlocking systems of air, water and minerals nourishing the earth.

and called for:

expanded co-operation and greater commitment to basic studies involving many individual scientists and new ventures such as the World Climate Program and will call for an integrated effort of many disciplines, including scientists from all regions of the world. Continuation of efforts at the present level will be too little, too late.

Accordingly we draw attention to the fundamental scientific importance of understanding the biogeochemical cycles which link and unify the major chemical and biological processes of the earth's surface and the atmosphere.

In 1983 the Executive Board of ICSU, in its meeting in Stockholm, for the first time considered that:[5]

a central intellectual challenge of the next few decades is to deepen and strengthen our understanding of the interaction among the several parts of the geosphere and biosphere, including the anthropogenic impact on biological productivity.

This meant implicitly that the issue was broader than climate, and that ICSU could do more than it did in WCRP. It also reaffirmed that SCOPE was not designed as a research programme, and thus was not equipped to carry out the research priorities it had identified during its work. Also pragmatic considerations like the great potential of newly available computer and satellite technology for addressing global problems played a role.

Later in 1983 the ICSU General Committee discussed in Warsaw the issue of a major research programme on the interaction of biological, chemical, and physical systems of the global environment and decided:

to establish an *ad hoc* committee to carry out the study of those aspects of global change which are not yet adequately covered, to prepare an inventory of existing programmes and their inter-relations, and to make recommendations to the 20th General Assembly for further planning in fields that are interdisciplinary and require international cooperation.

The outcome was a symposium at the General Assembly in Ottawa in 1984, organised by T F Malone and J G Roederer.[6] The programme comprised lectures on a diversity of global change issues such as R Revelle's paper on 'Soil dynamics and sustainable carrying capacity of the Earth'. It concluded with a summary by W S Fyfe (Canada):[7]

We live in a special period of the history of our planet. Man, because of his intelligence and success, is changing the environment, and has become the major force in the transport of solid earth materials, and his chemical by-products are changing the hydrosphere and the atmosphere. At the same time, this intelligence has developed the science and technology to observe and record change. We can observe on virtually every scale: from sub-atomic particles, to the arrangement of atoms and molecules, to the dynamics of the planet.

Fyfe referred to the growing availability of data and to the advances in technique, not only observation from space, which is very striking, but also the vastly improved techniques of observation at the microscopic and atomic levels which play a part in the study of the interactions of which the outcome is seen in massive changes. The whole range of scientific study is thus involved, not only that of the visible and obvious.

Fyfe put a key issue:

> The present period of accelerated population growth must cease in the next century or so; however, most would agree that human population will reach 10 billion by next century. It is imperative that we monitor response to this increase, to provide the background knowledge for intelligent planning, maintenance and, it is to be hoped, improvement of the human situation.

The General Assembly voted to set up an *ad hoc* planning group under B. Bolin (Sweden) which reported to the next General Assembly (Berne, 1986) where it was decided to initiate the 'IGBP *a* study of Global Change', a title indicating that not all aspects of global change would be taken on board. In particular it was agreed in the discussion in Berne that issues belonging to the domain of the social sciences would not be included, as the task for the natural sciences was already heavy and complex enough. A Special Committee for the IGBP (later Scientific Committee) was set up. It was chaired by J J McCarthy (USA); T Rosswall (Sweden) was the Executive Director. The objectives were to describe and understand:

(a) the interactive physical, chemical and biological processes that regulate the total earth system;
(b) the unique environment that this system provides for life;
(c) the changes occurring in this system;
(d) the manner in which these changes are influenced by human activities.

Within these objectives IGBP restricted itself to:[8]

> those issues that are deemed to be of greatest importance in contributing to our understanding of the changing nature of the global environment on time scales of decades to centuries, that most affect the biosphere, and that are most susceptible to human perturbations and that will most likely lead to practical capability.

The following core projects with a lifetime of about a decade were selected:

> How is the chemistry of the global atmosphere regulated, and what is the role of biological processes in producing and consuming trace gases? (International Global Atmospheric Chemistry Project, IGAC.) Priority issues: chemical transformations, and biospheric sources and sinks (marine and terrestrial), of atmospheric constituents that have a role in controlling the global system, especially radioactive active trace gases, aerosols, and reactive radicals.

> How will global changes affect terrestrial ecosystems? (Global Change and Terrestrial Ecosystems, GCTE.) Priority issues: responses of natural and managed ecosystems to changes in climate, atmospheric composition and land use, with emphasis on both impacts and feedback processes, investigated experimentally and through the development of dynamic vegetation models; the role of ecological complexity in the functioning of the global system.

> How does vegetation interact with the physical processes of the hydrological cycle? (BAHC, Biospheric Aspects of the Hydrological Cycle.) Priority issues: the effect of land surface properties (soil, vegetation, and topography) on water, carbon, and

energy fluxes, temporal and spatial integration of these processes, including improved modelling of complex landscapes; the down-scaling of climate information obtained from general circulation models.

How will changes in land-use, sea-level rise, and climate alter coastal ecosystems, and what are the wider consequences? (LOICZ, Land–Ocean Interactions in the Coastal Zone.) Priority issues: the effects of changes in external forcing on coastal fluxes; coastal biogeomorphology and sea-level rise; carbon fluxes and trace gases emissions; economic and social impacts of global change on coastal systems.

How do ocean biogeochemical processes influence and respond to climate change? (JGOFS, Joint Global Ocean Flux Study.) Priority issues: processes controlling the fluxes of carbon and associated biogenic elements within the ocean, and their exchanges with the atmosphere, sea floor, and continental boundaries; interpretation and application of remotely sensed ocean colour data.

What significant climatic and environmental changes occurred in the past, and what were their causes? (PAGES, Past Global Changes.) Priority issues: high resolution reconstruction, at the global scale, of the changes occurring in the past 2,000 years; investigations of the more radical re-organisations of the global system during the most recent glacial/interglacial cycles.

Two other Core Projects were developed for possible acceptance: Land Use/Cover Change (LUCC), and, at an earlier stage of planning, the Global Ocean Euphotic Zone Study (GOEZS).[9]

In addition, three activities of an overarching nature were now operational:

A Task Force on Global Analysis, Interpretation, and Modelling (GAIM), with responsibility, evaluation, and application of comprehensive prognostic models of the global biogeochemical system, and subsequently linking such models to those of the physical climate system.

The IGBP Data and Information System (IGBP-DIS), with responsibility for assisting Core Projects in meeting their data acquisition and data-management needs; also facilitating collaboration with the space agencies, other data-producing bodies, and international data centres.

The Global Change System for Analysis, Research, and Training (START), that promotes regional capacity building in global change science, and the establishment of networks for regionally based research and analysis relevant to the origins and impacts of global environmental change.

They were presented in detail in IGBP report no. 12 to the Scientific Advisory Council (La Villette, Paris, 1990) which gave its endorsement, thus marking the beginning of the operational phase and the end of the planning phase which had comprised 50 planning meetings in which close to 500 scientists participated.

During this planning phase the planners had successfully resisted pressures to broaden the programme. The last thing they wanted was the creation of an incoherent umbrella programme. But there were complaints that there 'was not enough biology, geology, soil science, etc. in the programme' and several Unions and Committees who were already working on global change issues felt their efforts should be connected to IGBP. Also the WCRP and SCOPE had to determine their position *vis à vis* IGBP, as their mandates bordered on that of the new programme.

These developments were among the reasons why the Executive Board of ICSU had created the Advisory Committee on the Environment (ACE)[10] in 1989 for the purpose of:

> providing the Board with advice on the status and development of all ICSU bodies' activities related to the environment and Global change in order to:
>
> promote co-ordination of efforts among members of the ICSU family;
>
> provide for a platform of discussion and exchange of information among relevant ICSU bodies and their external partners;
>
> set up and maintain an interface with external partners, such as UN bodies, and other international organizations including those in the social and engineering sciences and industry.

One of the first actions of ACE was to organise a one day Forum on Global Change in the framework of the IGBP SAC meeting at la Villette. In this Forum ICSU bodies were invited to make presentations on work relevant to Global Change and not included in the focussed research programme that it was hoped IGBP would be. This not only provided for visibility and recognition of this work, but, more importantly, created opportunities for practical liaison at working level between IGBP and ICSU bodies. By (1994) 26 liaison persons with ICSU bodies had been appointed.

One of the speakers of this Forum was Maurice Strong,[11] Secretary General of the forthcoming United Nations Rio Conference. Strong had extensive experience in managing international conferences, particularly when connected with environmental matters. In his speech he invited ICSU to act as principal science adviser to the preparatory process for the conference which was accepted by the President of ICSU. Together with COSTED, ACE was to oversee ICSU's activities in this advisory role.

IGBP largely followed the formula of GARP and WCRP: national scientific members of ICSU were asked to appoint national IGBP committees (69 by 1994) and in a consultation between these committees and the Scientific Committee for IGBP a framework research plan was developed. Into this the national global change research activities could be dovetailed, obviously to great advantage in terms of synergy and economy of manpower and costly equipment. Countries contributed in two ways: by direct annual payments for the central planning and co-ordination work, and in kind, by hosting and staffing a secretariat or core project office. Most important, of course, was their own research contribution.

The response was satisfactory with regard to the research and in contributions in kind, but during its lifetime IGBP had to struggle with a shortage of funds for central planning and co-ordination (so called 'core funding') because donors preferred to fund concrete, finite projects. As Global Change Research grew in importance, both nationally and internationally, an informal group of Funding Agencies for Global Change Research (IGFA[12]) was formed, following the initiative of R W Corell (US National Science Foundation). Its activities included to

> exchange information and national global change research programmes, supporting programmes and facilities;
>
> discuss approaches to the integration and phasing of implementation of global change research in light of available resources;
>
> promote coordination of access to and deployment of specialised research facilities; and
>
> aim to optimise allocation of national contributions to global change research.

ICSU, IGBP, WCRP, and HDP were usually invited to attend IGFA annual meetings as observers. Understandably, core funding was regularly on the agenda. In 1992 the IGBP central secretariat budget was about 4 million US $, while the total IGBP relevant research effort was estimated by IGFA at over 600 million $ in that year. Undoubtedly IGBP was becoming the largest and most complex international interdisciplinary research project so far undertaken. It was recognised along with WCRP by the General Assembly of the UN and relevant UN agencies. The OECD, which had been assessing the distribution of world scientific requirements, declared Global Change Research in 1993 a 'distributed' kind of Megascience in contrast to the existing Megascience topics, all of which concentrated on the use of one expensive piece of equipment like an accelerator or an astronomical observatory.

ASCEND 21, UNCED and the follow-up process

Rio (UNCED) marked the transition between a preparatory process and an implementation process of the results of the conference. Many organisations that felt they could contribute to the follow-up process of the results of UNCED made efforts to participate in the preparations to help shape these results. This was different for ICSU, which had on its own taken up the study of the Earth System through SCOPE, WCRP, and IGBP, and had concerned itself with development through COSTED since 1966. Also, even without the request to act as science adviser to UNCED, ICSU most probably would have taken the initiative to convene its conference on 'Agenda of Science for Environment and Development into the 21st Century' (ASCEND 21), but in the circumstances this meeting was able to serve two purposes, ICSU's own as well as that of providing advice to UNCED.

This comprised assistance in vetting drafts of Conference documents on scientific quality and proposing names of scientists who could help in the drafting of specific chapters of AGENDA 21, UNCED's action plan. Of particular importance was *Chapter 35*: 'Science for sustainable development', for which ICSU provided substantial inputs. These were, however, not always accepted because of the political consensus that had to be achieved at the Conference. Working on the draft and preparing for ASCEND went hand in hand, and in the end *Chapter 35* could state that its content was 'in harmony' with the outcome of ASCEND 21.

ASCEND, a unique Conference, organised within one year, was a critical survey by scientists, physical, chemical, biological, medical, and social, with engineers as well, of the factors relevant to the global environment and the management of 'sustainable development'. So various a meeting needed and received the co-sponsorship of other bodies: (TWAS, IIASA, the European Science Foundation, the Norwegian Research Council for Science and the Humanities, the Norwegian Academy of Science and Letters, the Stockholm Environment Institute, and the International Social Science Council).

There were 300 invited participants from over 70 countries, half of them from developing countries. Everyone who came did so on a personal basis. Papers on 16 themes were prepared in advance by expert teams (most often representing both advanced and developing countries). Discussion and revision by the Conference provided matter for a final publication in May 1992.[13]

The introduction to the publication declares:

> The objectives of the ASCEND 21 Conference were to bring together the understanding and judgment of the world scientific community (encompassing natural, social, engineering and health scientists) on the issues of highest priority for the future of the environment and development, so as to define the agenda of science in these areas for the next 10 to 20 years. The outcome was to provide both a consolidated contribution to the Rio Conference and a perspective for the future of international science in these and related areas.

Such a preamble tells us one important thing: ICSU had undergone a conversion (or progression) to a body possessed of a unity it never quite achieved when it evolved out of IRC. It goes too far to say that ICSU had recovered something it lost in 1931. IRC had a unity of authority only on paper. ICSU had made no claim to authority except by the consent of members specialised in particular topics. But, with the publication of ASCEND 21, ICSU showed that it could be concerned, on behalf of all its members, with a single major purpose. It could be so concerned only at times: each Union is concerned with its subject (the chemists with chemistry, the physicists with physics), each national member with its own national interests or specialities (as, for example, the French with science studied in French territories). Each has its own physical or intellectual territory with which it alone is competent to deal.

Why emphasise this? Was not global thinking already present in much of the work of ICSU and its members? It was, but with ASCEND it moved from the incidental into the general. One cannot quite say 'dominant' because all the Unions and all of those associate bodies which were not specifically created for global projects still kept most of their energy for manageable projects on a deliberately limited scale. While biologists, for example, easily turn to global environmental issues, crystallographers do not have to think in world terms (unless they happen to be molecular biologists as well). What ICSU had been able to do for a long time was to bring members together in new units with wide remits. The thinking which led to ASCEND led also to ICSU imposing on itself a permanent obligation to equip itself to act in global terms.

This was particularly evident from the final statement made by President Menon in his closing address:

> In this regard I would like to make a formal statement which will outline ICSU's plans.
>
> The contributions of science to environment and development have been rapidly growing over the past few years and so have the expectations of those responsible with respect to the assistance science can provide in their policy making programs. Consequently, and in accordance with its responsibility in the scientific community, ICSU has decided to take the following steps as a follow-up to both ASCEND 21 and UNCED:
>
> (a) to further consolidate the cooperation and coherence of ICSU's major international research programs and to promote their collaboration with relevant programs outside ICSU with special attention to be given to co-operation with the social sciences;
>
> (b) to strengthen further its capacity to play its role in the evolving partnership among science, governments, intergovernmental organizations and business and industry;
>
> (c) to strengthen the capacity of ICSU to prepare objective scientific assessments of issues of environment and development within its competence;
>
> (d) to enhance the efforts of ICSU to report scientific issues to the general public as well as to decision-makers in a readily understandable manner;
>
> (e) to strengthen its program for capacity building in science in developing countries;
>
> (f) to declare its willingness to play a role in helping whatever mechanism is set up to review Agenda 21 performance, e.g. by evaluating adequacy of scientific inputs or by helping define indicators by which sustainability and progress towards it can be measured.

The ASCEND meeting had discussed the following themes:

> Problems of environment and Development.
> – Population and Natural Resources Use
> – Agriculture, Land Use and Degradation
> – Industry and Waste
> – Energy
> – Health
>
> Scientific Understanding of the Earth System.
> – Global Cycles
> – Atmosphere and Climate
> – Marine and Coastal Systems
> – Terrestrial Systems
> – Freshwater Resources
> – Biodiversity
>
> Responses and Strategies.
> – Quality of Life
> – Public Awareness, Science and the Environment

– Capacity Building
– Policies for Technology
– Institutional Arrangements

At first reading this seems unremarkable: a list of topics already engaging the attention of many ICSU bodies. However it is the conditions under which they were gathered together and their enunciation as common policy that is significant. These were summarising continuing interests not only of separate ICSU bodies, but of the whole ICSU family. ICSU membership had seemed very scattered in the period before the decade (1956–66) which saw the mounting of IGY and the foundation of COSTED. It then began to coalesce with the planning of those two kinds of function (one a research programme, the other a widespread service) and the trend to convergence they exemplify.

Some of the participants in ASCEND 21 wrote with high hope afterwards that it had expressed a commitment on the part of the whole scientific community to work together, not only in research, but in the appropriate assessment of results. They looked forward to the effective prediction of impacts of new scientific discovery and technological innovation. They were careful not to be too confident that they would be listened to: they did no more than offer a model of a *modus operandi* for formulating policy options, to be offered to those authorities which bore political and economic responsibility as the consensus of international scientific opinion. We shall have to wait and see.

What was valuable in ASCEND 21 was that it was a cross-section, as it were, of the state of thought of informed scientific opinion about many world problems at one time. It was important that all the papers should bring their statements up to the minute. ICSU, as adviser to UNCED, might therefore claim to have taken an Olympian view of science as it concerned the human condition. The earliest enthusiasts for international science at the beginning of the century, seeking little more than friendly communication, would have been astonished at such pretension, but the organisers of ASCEND 21 had no misgivings. Although each specialist kept to his or her familiar field, each one spoke of a concern for the world as a whole.

Each also had his or her own ambitions, but between them they identified some areas of common interest which deserved specific action:

> intensified research into natural and anthropogenic forces and their inter-relationships, including the carrying capacity of the Earth and ways to slow population growth and reduce overconsumption;
>
> strengthening support for international global environmental research and observations of the total Earth system;
>
> research and studies at the local and regional scale on: the hydrological cycle; impacts of climate change; coastal zones; loss of biodiversity; vulnerability of fragile ecosystems; impacts of changing land use, of waste, and of human attitudes and behaviour;
>
> research on transition to more efficient energy supply and use of materials and natural resources;
>
> special efforts in education and in building up of scientific institutions as well as involvement of a wide segment of the population in environment and development problem-solving;

regular appraisals of the most urgent problems of the environment and development and communication with policy-makers, the media, and the public;

establishment of a forum to link scientists and development agencies along with a strengthened partnership with organisations charged with addressing problems of environment and development;

a wider view of environmental ethics.

The views of ASCEND 21 thus provided the basis for *Chapter 35* of UNCED's publication AGENDA 21: 'Science for sustainable development'.

There were 1,420 NGOs accredited to UNCED. ICSU could consider itself outstanding among them, partly through its special standing with UNESCO, partly through having been one of only 12 invited to select a representative to address a plenary session.

ASCEND was one thing: a meeting of scientists with a common habit of mind. UNCED was quite another: it was a governmental conference involving political power and national interests. One hundred and seventy-six countries were represented; 107 Heads of State attended. There were 2,500 delegates and 800 media persons. The 1,420 NGO's that were accredited to UNCED set up their own 'Global Forum' with information booths and discussion tents. Since most of these were advocacy groups, ICSU had a special position among them.

Although many of the participants and those reading about it in the newspapers were disappointed, UNCED has marked a historic turning-point: for the first time so many governments had come together to address the evidence put before them by science about the limitations of the planet to support the development of the totality of their collective countries and to initiate action to safeguard the 'common future'. And this they did by signing the action plan *Agenda 21* and agreeing two conventions. UNCED also reaffirmed that science had come to stay as a necessary ingredient in global policy-making.

The President of ICSU, Professor M G K Menon, in addressing the Plenary Session of UNCED, referred to the long history of limited global studies that the scientific community had carried out for over a century, but said also that this did not represent the totality of efforts needed to bring about sustainable development.

> Scientists are aware that for this, while science and technology are important, societal transformation is paramount. There are many ways in which we can mitigate adverse aspects of global change even today – and contribute to significant removal of poverty. But this is not happening. The problem lies in the crucial sphere of human relations. Scientists realize that, apart from advancing science and its applications that are appropriate for sustainable development, science must forge partnerships with other sectors of society; with social and engineering sciences, with business and industry, with governments and intergovernmental organizations, and, very important, the independent voluntary non-governmental sectors which work directly with society. The scientific community recognizes that it is part of society and must contribute to the needed societal transformations that bring about sustainable development.[14]

Although only a small amount of funds had been pledged at Rio for implementation, the follow-up process started immediately after, at governmental and non-governmental levels. For instance, the two conventions were ratified before the end of 1994, the UN created the Commission on Sustainable Development (CSD) which was to oversee the execution of Agenda 21, WMO held a special intergovernmental meeting to invigorate the World Climate Programme in the light of UNCED, China developed its own Agenda on the basis of that of UNCED, the Earth Council was set up to follow, from the non-governmental side, the implementation of Agenda 21.

What did ICSU do to follow up ASCEND and UNCED? In part it continued activities and initiatives started before 1992, but also took new steps. Liaison was made with the CSD and the Earth Council as well as with the Secretariats of the Convention for Climate Change and that for Biodiversity. Already in 1991 IUBS, SCOPE, and MAB of UNESCO had started an international programme called DIVERSITAS,[15] which comprised three themes: the role played by biodiversity in ecosystems, including the role of 'cornerstone' organisms, the disappearance of which would have profound effects on ecosystem functioning; the origins, maintenance, and loss of biodiversity; and the problems associated in inventorying and monitoring biodiversity, complicated by the lack of competent taxonomists, themselves deemed to be an 'endangered species'. The programme could well interact with the Convention on Biodiversity. Similarly IGBP and WCRP could do the same in respect of the scientific requirements of the Framework Convention on Climate Change.

Around 1992 another important development took place in which ICSU was a partner with several UN bodies. It was being recognised that observations of the atmosphere, the oceans, and the land surface had been done piecemeal and for a variety of limited purposes, rather than in a holistic manner such as would aim at keeping watch over the general state of the globe. Thus planning started for a Global Climate Observing System (GCOS), followed by a Global Ocean Observing System (GOOS) and, somewhat later, by a Global Terrestrial Observing System (GTOS). All would make use of existing observing systems and complement and update them where necessary. ICSU is a partner in the planning phase, since science is not only needed for the technical design to the observing systems, but is also an important client of the results. Mechanisms for co-operation between the three systems have been put in place.

Besides these specific activities, many members of the ICSU family had already embarked on or were initiating actions that could be considered as implementing ASCEND 21 and/or AGENDA 21. In the area of capacity building, as pledged in ASCEND, the Executive Board appointed in 1993 a Committee on Capacity Building in Science. The START initiative of IGBP, WRCP, and the Human Dimensions Programme made considerable progress in establishing networks in the Third World that would *inter alia* help create capability in Earth System Science. In 1993 also, COSTED and the International Biosciences Networks joined forces to form COSTED/IBN which, with the support of UNESCO, promised to be a valuable instrument in Capacity Building in science and technology in developing countries.

In the areas of global change, the work of ICSU bodies like SCOR, SCAR, SCOPE,

SCOSTEP, SC-IDNDR, CODATA, COSPAR, and IUPAC continued, and some of these made presentations at the Forum on Earth System Science organised by ACE in the framework of the IGBP/SAC meeting at Ensenada in January 1993. On this occasion J C I Dooge, President-elect of ICSU gave a personal evaluation of ICSU's response to ASCEND over the previous year. He registered progress in many areas, notably in Earth System Science. However, in creating public awareness and links with industry, no success could be recorded.

In addition, the links with the engineering and social sciences, despite a successful meeting with the relevant partner NGOs, had remained weak. However, as IGBP became operational and began to produce results, it had become strong enough to widen its scope and engage in collaboration with the social sciences when this was useful. Because of the decision in Berne (1986) not to include social science aspects in IGBP, the International Social Science Council (ISCC) had initiated, in 1990, the development of the Human Dimensions of Global Environmental Change Programme (HDP) which led in 1992 to the presentation of a work plan. Although governments had high expectations of the social-science contribution, it proved difficult to bring HDP into operation, to some extent because ISSC had yet no tradition of major international research programmes, nor did it possess a strong constituency of national scientific members. Since both ICSU and ISSC felt that collaboration would be to their mutual advantage, interaction at various levels was encouraged, as exemplified by the above-mentioned projects. It should also be noted that, since 1993, out of the 23 ICSU Unions, 4 belong to the social sciences, pointing perhaps to a reconvergence process among the sciences stimulated by the need jointly to address large world problems.

In making up the balance, it must be concluded that ICSU has already taken several important steps along the global pathway it set for the scientific community in ASCEND 21, for which Ringberg and Visegrad had been a prelude. But, at the same time, ICSU has the tasks of fostering the scientific endeavour, in its ever self-renewing manifestations, and must look forward to future opportunities and horizons as offered, for instance, by the electronic superhighway. Also it should be realised that ICSU's outreach towards global problems is not equally relevant to all branches of science, and as a result some Unions feel more closely associated with it than others.

Chapter 19

ICSU at the end of a century

Look back. We have seen how ICSU began as a means of bringing together on an international basis two kinds of communities of scientists. There were those who had established international bodies of their own, each keeping, on the whole, to a single discipline, some disciplines, like chemistry, being centuries old (e.g. IUPAC), some having originated within a few recent decades (e.g. URSI). There were those bodies, generally classified as academies, who represented the scientific community of all disciplines within a single nation. The objective in bringing these bodies together was both high-minded and practical: to act together for the intellectual good of mankind by unifying the practice of scientific research. Initially the industrial and economic consequences of science were not ICSU's concern, and no effort was made to work out any guiding philosophy of relations with the world of manufacture and money. Gradually however, it had become apparent that, if ICSU was to have a continuing influence in scientific research and its applications, it had to consider science in the world at large, the training of the next generation, and responsibility outside the lands which had inherited the European tradition. Moreover, while the traditional science had observed and measured a world which appeared fixed and separate from human involvement, mid 20th-century science was building a bridge of understanding between mankind and the external physical world as viewed by the scholar. From the time of the ICSU Presidencies of Thompson and Harrison in the 1960s to the recent period of J C Kendrew and M G K Menon we can follow changes in ICSU which were continuous and progressive.

Should we think of ICSU as introvert, extrovert, or both? Compare the background to two Presidential addresses, that of Harold Thompson in 1966 and M G K Menon in 1993 . Each was coming to the end of a term of office. The time span is roughly one generation. A scientist born during Thompson's Presidency would have reached full career during Menon's.

Thompson spoke in Bombay. He had to speak twice, once briefly to welcome delegates before giving the floor to the principal Indian host. As has already been pointed out, this was the first time the General Assembly had been held outside the western world, but he put no great emphasis on this, leaving it to Professor H J Bhabha (who was to die in a plane crash on Mont Blanc a few weeks later). Thompson's address opening the official business of the General Assembly was inward-looking. It could have been carried out anywhere. He was concerned with the administrative structure of ICSU, with the success of its scientific work, constitutional weaknesses under repair, uncertainty of relations with UNESCO.

The 1993 General Assembly opened with speeches of welcome from representatives of the Chilean government and of Chilean science, both of whom made the vital point that this was the first General Assembly of ICSU in South America. Menon was able to take this up, but moved quickly on to take a world view of ICSU's functions. He was speaking to a hall filled by a large gathering of Chilean scientists as well as the ICSU participants, so it was appropriate that he should take a broad view, but all the same we can now see a marked difference from Thompson's scope. Menon had been at the UN 'Earth Summit' in Rio de Janeiro, and it was evidently still much on his mind. The scope of his speech was superficially much the same as any President's must be, like Thompson's, but it was different in that he spoke in a different way of ICSU's place in the world endeavours to benefit mankind. To Thompson, such benefits were a happy consequence of scientific work: to Menon they were a driving purpose.

These differences were due to some extent to differences in personality, but they were also due to ICSU itself having changed. Can we speak of ICSU itself having a personality? If so, can we say that it was once introvert and that it has changed to extrovert? There is something in this: the scientists who formed IAA and IRC were mostly interested in the academic practice of scientific inquiry, even if their material was sometimes (as with IUGG and URSI) to be found in the external world. ICSU did not change this, but the decade around the Second World War made everyone look outwards.

We have seen changes exemplified in the Antarctic Treaty and in the formation of COSTED, in the broadening of the intellectual base of Union organisation, the enlargement of national membership, the creation of new modes of co-operation. These were the beginning of the extrovert outlook which, 30 years later, unites the many new organisations and programmes which we now see in ICSU's global conspectus. In this last decade of a century in which science has become something different from the science of previous centuries, ICSU has become aware of both its stature and its limitations. Science cannot do everything mankind needs, but it can do more and more. Everyone is entitled to his life, the farmer to his farming, the musician to his music, the artist to his art, the public servant to pride in the efficient dispatch of business. None should think that what he does is better for mankind than any other calling. However, the scientist does have a unique concern: the comprehension of the material world in which each of us must pass his life. In the past generation we have come to appreciate the extent to which mankind is involved in the mechanism of that material world, physically and biologically, and to appreciate how much mankind thereby can sustain both profit and loss.

As Menon implied, ICSU was not the only body with scientific interests involved in UNCED, before, during, and after the Rio meeting. The engineers had their own programme to launch during UNCED: a *World Partnership for Sustainable Development*. Behind this were WFEO (World Federation of Engineering Organisations), FIDIC (International Federation of Consulting Engineers) and CIESIN (Consortium for

International Earth Science Information Network). TWAS had an obvious involvement, together with TWNSO (Third World Network of Scientific Organizations) which began its own Programme for Sustainable Development.

Co-operation of this kind was not confined to the groups with particular subject interests: four national academies (India, Sweden, UK, USA) soon held a conference on World Population in Delhi in October 1993.

At Rio, UNCED was a governmental conference. Those who saw it as a great scientific opportunity had some disappointments. However, considering the amount of scientific input, it might have been a worse conference from the point of view of future action. Within the existing international non-governmental science–technology community, much was learned about communicating to the political sphere. Perhaps the most important lesson was the need to be early, persistent, and thorough. Anyone who has been any length of time in the government service in any country will have learned that there is a language to be acquired. If there are differences in each separate country, it only means that civil servants in one country use a dialect of a common language of administration and decision-making, which they share with their opposite numbers in other countries. This is not the language of research or of commerce: it is a language which must be learned, and those who are not of this calling may have to be content to communicate in it without fluency, and to eavesdrop in the corridors of power.

In idealist words: the scientist (for which one can often read ICSU) can no longer be content to be driven only by curiosity or by the ambition to create new material benefits in manufacture, agriculture, defence (so called) or in medicine; he must be driven also by the universal need for that development without which there can be no ultimate survival.

Idealism, however, is all very well: what is claimed to be practical politics sometimes takes over to disconcert the idealist. UNESCO seemed to many the embodiment of high aspiration, but there can be different views of how it is in fact operating. When the United States government withdrew its membership, to be followed soon after by the United Kingdom government, the loss of subscription revenue to UNESCO was serious. However it continued to function. Both the USA and the UK continued to be active members of many bodies associated with UNESCO so long as they were not directly involved in UNESCO's political governmental structure. The fact that ICSU, like many other NGOs, and its Unions continued to enjoy the support of the USA and the UK highlights the importance of the nature of national membership of ICSU: it is the representative body of the scientific community of a country which is the member, not its government.

Many of them are called *academy*. How tidy it would all be if the word *academy* always meant the same thing. But it does not. The range of character of representative national bodies is wider even than that of the Unions. The reasons are historical, political, and scholarly.

One can make some unreliable generalisations. Historically one sees a paradox in that some of the most independent academies are those which were formed at the wish of, or with the approval of, monarchs. Some keep 'Royal' in their titles or have lost it in the process of political change. Some academies have been formed by scientists themselves and have obtained government approval. Some academies are creations of governments exercising strong central control who desire that the international face of their scientific activity should be respectable. Some have been formed by the pressure of a scientific community anxious that science should have an independent voice in government circles.

Politically the academy idea is attractive: a group of like-minded scholars meeting for mutual enlightenment without ties other than familiarity with an agreed meeting-place. The idea is admirable and out of date. The modern academy must look for some power or authority, financial means, or political mandate, to exert influence on behalf of its members for the prosecution of their studies. All the more reason, therefore, for any academy to look to others for moral support and for common action.

The scholarly function of the academy is the one that has endured over the centuries, but this takes on different aspects as the range of scientific inquiry expands. An academy cannot take every scientist into membership. Excellence must be the criterion, but as time goes by new kinds of excellence appear as new sciences emerge. This is why specialist learned and professional bodies were formed one by one in the 19th century and have proliferated in the 20th. (Dissatisfaction with the old academies was also a factor, and it is by no means certain that all can survive without radical change.)

These generalisations would be useful if we were only to think about the older countries of the Northern world, but we cannot extend them to countries which have changed or been changed in recent times (as in many South American countries), or have deep-rooted cultures which have adapted to change (the outstanding example being India). Science is not what it was, North or South. What is the difference between modern conditions and those in which the International Association of Academies was formed nearly a 100 years ago or IRC nearly 80? To answer this one would need to write half a dozen histories including the kind of history of technology, which is expressed in the title of a distinguished periodical *Technology and Culture*. It is easy to assert that 'we' are all much more conscious of the impact of applied science on our lives than were our great-grandparents, and that we sometimes act on that knowledge. 'We' in this context has to mean the kind of people who may read this book, and books like it, which means a minority of the world's enormously expanded population. To say this is not conceit, but an acceptance of the fact that being critically aware of the influence of science and technology in human life is still the good fortune of a few. But it is now many more than it was a hundred years ago.

It is a paradox that the Unions, which one might suppose to be more various than national academies, show more consistency in style and structure than many academies. The Unions are more like each other than the academies are like each other. The variety of academies arises to some extent from the differences in national funding of research. But there is an interaction. The increasing variety of Unions arises from the extension of scientific knowledge that is generated by the leaders who are personal members of the academies.

Money always came into it, but now more and more. Government funding of science is rarely high-minded or purely charitable. There has been some support for applied science from at least the 17th century. To give only a few prominent examples, there was support for astronomy as an aid to navigation, for analytical chemistry as an aid to taxation, and various aids to the military. The present century has seen the development of Ministries charged with the management of government funds for 'pure' research by universities and other semi-independent institutions, and departments within Ministries of Defence, Health, Transport, and others controlling research directed at their specific needs. No two countries seem to be the same in this respect, so that when those responsible for the finances of ICSU consider national membership they cannot think in simple terms of a scale of subscription categories. The formula for a scale may be simple: the actual expectations from some nations cannot.

To some extent ICSU is dependent on national membership for its maintenance. On the other hand there are nations which look to ICSU for support for the scientific work which their national budgets cannot or will not support, which is particularly pressing with the emergence of independent scientific communities in the Baltic republics and other similar nations.

It will be for governments to bring about the conditions in which the harvest of scientific endeavour can be fully gathered in. ICSU is a non-governmental organisation and it can only inform and advise. But there is hope that its intellectual authority will one day be matched by a moral authority, built up by devoted men and women like those who created and served IAA, IRC, and ICSU, a world-wide influence which began nearly a century ago with conversations between friends.

Appendix 1

Acronyms

ASCEND	Agenda of Science for Environment and Development into the 21st century
BAHC	Biospheric Aspects of the Hydrological Cycle
CAME	Conference of Allied Ministers of Education
CASAFA	Inter-Union Commission on Application of Science to Agriculture
CAST	China Association for Science and Technology
CCBS	Capacity Building in Science Committee
CERN	European Centre for Nuclear Research
CETEX	Contamination by Extra-terrestrial Exploration
CIESIN	Consortium for International Earth Science Network
CIOMS	Council for International Organizations of Medical Sciences
CISM	International Centre for Mechanical Sciences
CNRS	Centre National de Recherche Scientifique
COBIOTECH	Scientific Committee for Biotechnology
CODATA	Committee on Data for Science and Technology
COGENE	Scientific Committee on Genetic Experimentation
COMSCEE	Special Committee on Science in Central and Eastern Europe
COSPAR	Committee on Space Research
COSTED	Committee on Science and Technology in Developing Countries
COSTED/IBN	COSTED/International Biosciences Network
COWAR	Committee on Water Research
CSAGI	Comité Scientifique pour l' Année Géophysique
CSD	Commission on Sustainable Development
CTS	Inter-Union Committee on the Teaching of Science
DFG	Deutsche Forschungsgemeinschaft
DSIR	Department of Scientific and Industrial Research
ECOSOC	United Nations Economic and Social Council
EEC	European Economic Community
EMBO	European Molecular Biology Organization
ENUWAR	Environmental Consequences of Nuclear War
FAGS	Federation of Astronomical and Geophysical Data Analysis Service
FAO	Food and Agriculture Organization
FGGE	First GARP Global Experiment
FID	International Federation for Documentation
FIDIC	International Federation of Consulting Engineers
GAIM	Global Analysis Integration and Modelling
GARP	Global Atmosphere Research Programme
GATT	General Agreement on Trade and Tariffs
GCOS	Global Climate Observing System

GCTE	Global Change and Terrestrial Ecosystems
GEWEX	Global Energy and Watercycle Experiment
GOEZS	Global Ocean Euphotic Zone Study
GOOS	Global Ocean Observing System
GTOS	Global Terrestrial Observing System
HDP	Human Dimensions of Global Environmental Change
IAA	International Association of Academics
IAEA	International Atomic Energy Agency
IAM	International Association of Microbiologists
IAMS	International Association of Microbiological Societies
IAPSO	International Association of Physical Sciences of the Ocean
IAU	International Astronomical Union
IAVSD	IUTAM Vehicle System Dynamics
IBN	International Biosciences Networks
IBP	International Biological Programme
IBRO	International Brain Research Organization
IBY	International Biological Year
ICA	International Cartographic Association
ICASE	International Council of Associations of Science Education
ICES	International Council for the Exploration of the Sea
ICHPS	International Council for Philosophy and Humanistic Studies
ICOM	International Council of Museums
ICRO	International Cell Research Organization
ICSTI	International Council for Scientific and Technical Information
ICSU AB	ICSU Abstracting Board
ICSU	International Council of Scientific Unions
ICTP	International Centre for Theoretical Physics
IDNDR	International Decade for Natural Disaster Reduction
IGAC	International Global Atmospheric Chemistry Project
IGBP–DIS	IGBP Data and Information System
IGBP	International Geosphere–Biosphere System
IGFA	International Group of Funding Agencies for Global Change Research
IGU	International Geographical Union
IGY	International Geophysical Year
IHP	International Hydrological Programme
IIASA	International Institute for Applied Systems Analysis
IIC	International Institute for Intellectual Co-operation
IIOE	International Indian Ocean Expedition
ILO	International Labour Organization
IMA	International Meteorological Association
IMM	IUTAM Mechanical Mathematics
IMO	International Meteorological Organization
IMU	International Mathematical Union
INASP	International Network for the Availability of Scientific Publications
IOC	Intergovernmental Oceanographic Commission
IPCC	Intergovernmental Panel on Climate Change
IQSY	International Years of the Quiet Sun
IRC	International Research Council
IRPTC	International Register of Potentially Toxic Chemicals
ISCC	International Social Science Council

ISCM	International Society for Contemporary Music
ISIP	International Science and Its Partners
ISM	International Society for Microbiology
ISCC	International Social Science Council
ISSS	International Society for Soil Science
ITU	International Telecommunication Union
IUAES	International Union of Anthropological and Ethnological Sciences
IUB	International Union of Biochemistry
IUBMB	International Union of Biochemistry and Molecular Biology
IUBS	International Union of Biological Sciences
IUCAF	Inter-Union Commission on Frequency allocation
IUCN	International Union for Conservation of Nature
IUCr	International Union of Crystallography
IUFRO	International Union of Forestry Research Organizations
IUGG	International Union of Geodesy and Geophysics
IUGS	International Union of Geological Sciences
IUHPS	International Union of the History and Philosophy of Science
IUIS	International Union of Immunological Societies
IUMS	International Union of Microbiological Societies
IUNS	International Union of Nutritional Sciences
IUPAB	International Union of Pure and Applied Biophysics
IUPAC	International Union of Pure and Applied Chemistry
IUPAP	International Union of Pure and Applied Physics
IUPHAR	International Union of Pharmacology
IUPS	International Union of Physiological Sciences
IUPsyS	International Union of Psychological Science
IUTAM	International Union of Theoretical and Applied Mechanics
JGOFS	Joint Global Ocean Flux Study
JOC	Joint Organizing Committee
LOICZ	Land-Ocean Interactions in the Coastal Zone
LUCC	Land Use/Cover Change
MAB	Man and the Biosphere Programme
MIRCEN	Microbial Resource Centres
OECD	Organization for Economic Co-operation and Development
PAGES	Past Global Changes
PRC	People's Republic of China
PSA	Pacific Science Association
SCAR	Scientific Committee on Antarctic Research
SCFCS	Free Circulation of Scientists, Standing Committee
SCSPS	Safeguard of the Pursuit of Science
SC–IBP	Special Committee for the International Biological Programme
SCIGPB	Scientific Committe for the International Geosphere–Biosphere Programme
SCMSS	Standing Committee on Membership, Structure and Statutes
SCOPE	Scientific Committee on Problems of the Environment
SCOR	Scientific Committee on Oceanic Research
SCOSTEP	Scientific Committee on Solar–Terrrestrial Phenomena
SCOWAR	Scientific Committee on Water Research
SCSPS	Standing Committee for the Safeguard of the Pursuit of Science
START	Global Change System for Analysis, Research and Training
TNO	Toegepast Natuur-Wetenschappelijk Ondersoek

TOGA	Tropical Ocean and Global Atmospheric Study
TWAS	Third World Academy of Sciences
TWNSO	Third World Network of Scientific Organizations
TWOWS	Third World Organization of Women in Science
UATI	Union of International Technical Associations
UN	United Nations
UNCED	United Nations Conference on Environment and Development
UNDRO	United Nations Disaster Relief Organization
UNEP	United Nations Environment Programme
UNESCO	United Nations Educational, Scientific and Cultural Organization
URSI	Union Radio Scientifique Internationale
WCRP	World Climate Research Programme
WDC	World Data Centres
WFEO	World Federation of Engineering Organizations
WHO	World Health Organization
WIPO	World Intellectual Property Organization
WMO	World Meteorological Organisation
WOCE	World Ocean Circulation Experiment
ZWO	Zuiver Wetenschappelijk Ondsersoek

Appendix 2

Administration, management, history, finance

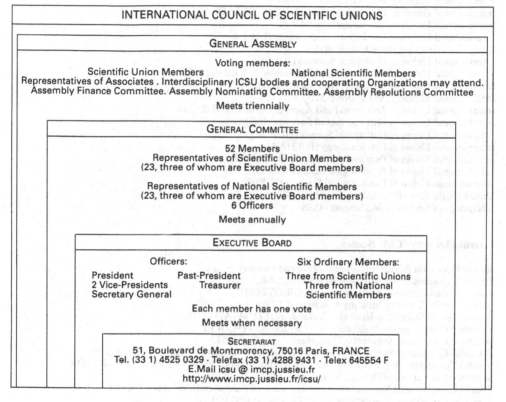

INTERNATIONAL COUNCIL OF SCIENTIFIC UNIONS

GENERAL ASSEMBLY

Voting members:
Scientific Union Members   National Scientific Members
Representatives of Associates . Interdisciplinary ICSU bodies and cooperating Organizations may attend.
Assembly Finance Committee. Assembly Nominating Committee. Assembly Resolutions Committee

Meets triennially

GENERAL COMMITTEE

52 Members
Representatives of Scientific Union Members
(23, three of whom are Executive Board members)

Representatives of National Scientific Members
(23, three of whom are Executive Board members)
6 Officers

Meets annually

EXECUTIVE BOARD

Officers: Six Ordinary Members:
President   Past-President   Three from Scientific Unions
2 Vice-Presidents   Treasurer   Three from National
Secretary General   Scientific Members

Each member has one vote

Meets when necessary

SECRETARIAT
51, Boulevard de Montmorency, 75016 Paris, FRANCE
Tel. (33 1) 4525 0329 · Telefax (33 1) 4288 9431 · Telex 645554 F
E.Mail icsu @ imcp.jussieu.fr
http://www.imcp.jussieu.fr/icsu/

Inter-Organisational Relations

ICSU has established relations with the following intergovernmental Organisations
United Nations Educational, Scientific and Cultural Organisation (UNESCO); Consultative and Associate
    Status (Category A) since 1961.
ICSU–UNESCO Coordinating Committee to ensure close cooperation in the implementation of scientific
    programmes.
World Meteorological Organization (WMO): Working arrangement since 1960;
International Atomic Energy Agency (IAEA): Consultative Status since 1960;
International Telecommunications Union (ITU): Working Agreement since 1962;
Food and Agricultural Organization (FAO): Specialized Consultative Status since 1963;
World Health Organization (WHO): Official Relations since 1964;
Economic and Social Council (ECOSOC): Category II Status since 1971;
United Nations Environment Programme (UNEP): Cooperation on informal basis since 1972;
Office of the United Nations Disaster Relief Coordinator (UNDRO) since 1980;
Council of Europe: Consultative since 1961;
United Nations International Decade for Natural Disaster Reduction (DNDR) since 1969;
Secretariat for UN Conference on Environment and Development (UNCED) and
Commission on Sustainable Development (CSD) since 1990;
World Intellectual Property Organization (WIPO) since 1993

Scientific Union Members

International Union of Anthropological and Ethnological Sciences (IUAES)
International Astronomical Union (IAU)
International Union of Biochemistry and Molecular Biology (IUBMB)
International Union of Biological Sciences (IUBS)
International Union of Pure and Applied Biophysics (IUPAB)
International Brain Research Organization (IBRO)
International Union of Pure and Applied Chemistry (IUPAC)
International Union of Crystallography (IUCr)
International Union of Geodesy and Geophysics (IUGG)
International Geographical Union (IGU)
International Union of Geological Sciences (IUGS)
International Union of the History and Philosophy of Sciences (IUHPS)
International Union of Immunological Societies (IUIS)
International Mathematical Union (IMU)
International Union of Theoretical and Applied Mechanics (IUTAM)
International Union of Microbiological Societies (IUMS)
International Union of Nutritional Sciences (IUNS)
International Union of Pharmacology (IUPHAR)
International Union of Pure and Applied Physics (IUPAP)
International Union of Physiological Sciences (IUPS)
International Union of Psychological Sciences (IUPsyS)
Union Radio Scientifique International (URSI)
International Society of Soil Science (ISSS)

Interdisciplinary ICSU Bodies

Ad hoc Group on Agriculture, Aquaculture and Forestry
Scientific Committee on Antarctic Research (SCAR)
Scientific Committee on Biotechnology (COBIOTECH)
Committee on Capacity Building in Science (CCBS)
Committee on Data for Science and Technology (CODATA)
Scientific Committee on Problems of the Environment (SCOPE)
Scientific Committee on Genetic Experimentation (COGENE)
Scientific Committee for the International Geosphere–Biosphere Programme (SC–IGBP)
Special Committee for the International Decade for Natural Disaster Reduction (SC–IDNDR)
Scientific Committee on Oceanic Research (SCOR)
Special Committee on Science in Central and Eastern Europe and the former Soviet Union (COMSCEE)
Scientific Committee on Solar–Terrestrial Physics (SCOSTEP)
Committee on Space Research (COSPAR)
Scientific Committee on Water Research (SCOWAR)
World Data Centre (WDC)
Federation of Astronomical and Geophysical Data Analysis Services (FAGS)
Inter-Union Commission on Frequency Allocations for Radio Astronomy and Space Science (IUCAF)
Inter-Union Commission on the Lithosphere (ICL)
Inter-Union Commission on Spectroscopy (IUCS)

International and Regional Scientific Associates

Academia de Ciencias de America Latina (ACAL)
International Institute for Applied Systems Analysis (IIASA)
Federation of Asian Scientific Academies and Societies (FASAS)
International Union Against Cancer (UICC)
International Cartographic Association (ICA)
International Cell Research Organization (ICRO)
International Dairy Federation (IDF)
International Federation of Societies for Electron Microscopy (IFSEM)
International Society of Endocrinology (ISE)

Engineering Committee on Oceanic Resources (ECOR)
International Union of Food Science and Technology (IUFoST)
International Union of Forestry Research Organizations (IUFRO)
International Foundation for Science (IFS)
International Federation for Information and Documentation (FID)
International Federation for Information Processing (IFIP)
International Council for Laboratory Animal Science (ICLAS)
International Federation of Library Associations and Institutions (IFLA)
International Life Sciences Institute (ILSI)
Pacific Science Association (PSA)
International Society for Photogrammetry and Remote Sensing (ISPRS)
International Union for Physical and Engineering Sciences in Medicine (IUPESM)
International Union for Quarternary Research (INQUA)
International Radiation Protection Association (IRPA)
International Federation of Scientific Editors (IFSE)
International Council for Scientific and Technical Information (ICSTI)
International Federation of Surveyors (FIG)
Third World Academy of Science (TWAS)
International Union of Toxicology (IUTOX)
International Union for Vacuum Science, Technique and Applications (IUVSTA)
International Association on Water Quality (IAWQ)

National* Scientific Members, Associates and Observers

*ICSU's National Scientific Members, Associates and Observers are institutions representing scientifically separate geographic areas. Country names are used for the sake of brevity and common usage and are not intended to carry political or diplomatic implications.

Argentina	India	Poland
Armenia	Indonesia	Portugal
Australia	Iran	Romania
Austria	Iraq	Russia
Bangladesh	Ireland	Saudia Arabia, Kingdom of
Belarus	Israel	Senegal
Belgium	Italy	Seychelles
Bolivia	Jamaica	Singapore
Brazil	Japan	Slovak Republic
Bulgaria	Jordan	South Africa
Burkina Faso	Kazakhstan	Spain
Canada	Kenya	Sri Lanka
Caribbean	Korea, Dem. People's Rep. of	Sudan, Republic of
Chile	Korea, Republic of	Swaziland
China: CAST	Latvia	Sweden
Taipei	Lebanon	Switzerland
Colombia	Lithuania	Thailand
Côte d'Ivoire	Madagascar	Togo
Croatia	Malaysia	Tunisia
Cuba	Mexico	Turkey
Czech Republic	Moldova	Uganda
Denmark	Monaco, Principauté de	Ukraine
Egypt, Arab Republic	Mongolia	United Kingdom
Estonia	Morocco	United States
Finland	Nepal	Uruguay
France	Netherlands	Uzbekistan
Georgia	New Zealand	Vatican City State
Germany	Nigeria	Venezuela
Ghana	Norway	Vietnam Socialist Republic
Greece	Pakistan	Zimbabwe
Guatemala	Panama	
Hungary	Philippines	

Joint Initiatives

Committee on Science and Technology in Developing Countries/International Biosciences Networks (COSTED/IBN)
International Geological Correlation Programme (IGCP)
World Climate Research Programme (WCRP)
Global Climate Observing System (GCOS)
Global Ocean Observing System (GOOS)
Global Terrestrial Observing System (GTOS)
Joint Lectureship/Professorship Programme
UNESCO/ICSU Short-Term Fellowships Programme

ICSU's Standing and *Ad hoc* Committees

Membership, Structure and Statutes (SCMSS)
Finance (SFC)
Freedom in the Conduct of Science (SCFCS)
ICSU Press
Advisory Committee on the Environment (ACE)

International Association of Academies (IAA)

Date of Creation

9 October 1899, Wiesbaden.

Initial Objectives

To initiate and otherwise to promote scientific understanding of general interest, proposed by one or more of the associated Academies, and to facilitate scientific intercourse between different countries.

Founding Members

Berlin
Göttingen
Leipzig
London
München
Paris
Rome
Saint-Petersburg
Vienna
Washington

International Research Council (IRC)

Date of Creation

July 1919, Brussels.

Objectives

To co-ordinate international efforts in the different branches of science and its application.
To initiate the formation of International Associations or Unions deemed to be useful to the progress of science, in accordance with Article 2 of the resolution adopted at the Conference of London, October 1918.
To direct international scientific activity in subjects which do not fall within the purview of any existing international associations.
To enter through the proper channels into relations with the Governments of the countries adhering to the IRC in order to promote investigations falling within the competence of the Council.

Founding Members

Australia	New Zealand
Belgium	Poland
Brazil	Portugal
Canada	Romania
France	Serbia
Greece	South Africa
Italy	United Kingdom
Japan	United States

International Council of Scientific Unions (ICSU)

Date of Creation

July 1931, Brussels.

Objectives

To co-ordinate the national adhering organisations, and also the various international Unions

To direct international scientific activity in subjects which do not fall within the purview of any existing international associations

To enter, through the national adhering organisations, into relations with the governments of the countries adhering to the Council in order to promote scientific investigations in these countries.

Members 1931

Unions

IAU
IUBS
IUPAC
IUGG
IGU
IMU
IUPAP
URSI

National members 1931

Argentina	Greece	Portugal
Australia	Hungary	Romania
Belgium	India	Siam
Brazil	Indochina	South Africa
Bulgaria	Japan	Spain
Canada	Latvia	Sweden
Chile	Mexico	Switzerland
Cuba	Monaco	Tunisia
Czechoslovakia	Morocco	United Kingdom
Denmark	Netherlands	United States
Dutch Indies	New Zealand	Uruguay
Egypt	Norway	Vatican City State
Finland	Peru	Yugoslavia
France	Poland	

General Assemblies of ICSU

GA	Place	Date	Under the Presidency of
1	Brussels	July 1931	E Picard
2	Brussels	July 1934	G E Hale
3	London	May 1937	N E Norlund
4	London	July 1946	H R Kruyt

GA	Place	Date	Under the Presidency of
5	Copenhagen	October 1949	J A Fleming
6	Amsterdam	October 1952	A von Muralt
7	Oslo	August 1955	B Lindblad
8	Washington, DC	October 1958	L V Berkner
9	London	September 1961	R Peters
10	Vienna	November 1963	S Horstadius
11	Bombay	January 1966	H W Thompson
12	Paris	September 1968	J M Harrison
13	Madrid	September 1970	V A Ambartsumian
14	Helsinki	September 1972	V A Ambartsumian
15	Istanbul	September 1974	J Coulomb
16	Washington, DC	October 1976	H Brown
17	Athens	September 1978	F B Straub
18	Amsterdam	September 1980	C de Jager
19	Cambridge	September 1982	D A Bekoe
20	Ottawa	September 1984	J C Kendrew
21	Berne	September 1986	J C Kendrew
22	Beijing	September 1988	J C Kendrew
23	Sofia	October 1990	M G K Menon
24	Santiago	October 1993	M G K Menon
25	Washington, DC	September 1996	J C I Dooge

Chronology of events in International Science Organization

Early Initiatives

1860	Karlsruhe Chemical Congress
1871	First International Geographical Congress
1877	International Bureau of Weights and Measures
	International Library Conference
1878	First International Geology Congress
1881	First International Congress of Electricity
1882–3	First International Polar Year
1889	First International Congress of Physiology
	Zoological Nomenclature
1897	International Committee on Atomic Weights
1897–9	Preparation for formation of IAA

The IAA and IRC Period

1899	Inaugural meeting of IAA in Wiesbaden
1900	International Commission on Photometry
1904	International Geological Congress wants IAA to support world programme in seismology
1906	Discussion of international organisation of meteorological stations
1907	International Polar Institute formed`
1909	International Committee for Publication of Annual Tables of Constants
1910	International Radium Standing Commission
1913	International Map of the World project discussed
1918	Conference on International Scientific Organisations
1919	IRC Inaugural meeting in Brussels
1919–31	Difficulties caused by post-War national sentiments. Preparations for transformation of IRC into ICSU

ICSU: The first 60 years

1931	ICSU founded. First General Assembly in Brussels
1932–3	Second International Polar Year
1934	Relations with Organisation for International Co-operation of League of Nations
1940–5	ICSU dormant during the Second World War
1947	Formal Agreement with UNESCO

1957–8	IGY
1962–7	Years of the Quiet Sun
1962–72	Upper Mantle Project
1964–74	IBP
1965–75	International Hydrological Decade
1966	COSTED started
1967–80	GARP
1968	MAB begins
1972	UN Conference on Human Environment, Stockholm
1973	IGCP started
1979	First World Climate Conference
	UNCSTD, Vienna
	IBN begins
1980	WCRP succeeds GARP
1985	Ringberg Meeting
	Villach Meeting
1986	IGBP started
1987	Brundtland Report
	Montreal Protocol on Ozone Layer
1988	IPCC formed
1990	Second World Climate Conference
	Visegrad Meeting
1991	ASCEND 21 Conference in Vienna
1992	UNCED in Rio de Janeiro
1992–4	GCOS, GOOS, and GTOS start planning
1996	First history of ICSU published

Past officers of ICSU

Elected	President	Secretary General	Treasurer
1931	G E Hale (USA)	H Lyons (UK)	
1934	N E Norlund (Denmark)	H Lyons (UK)	
1937	C Fabry (France)	F J M Stratton (UK)	
1946	J A Fleming (USA)	F J M Stratton (UK)	
1949	A von Muralt (Switzerland)	F J M Stratton (UK)	
1952	B Lindblad (Sweden)	A V Hill (UK)	W A Noyes (USA)
1955	L V Berkner (USA)	A V Hill (UK)	E Herbays (Belgium)
1956		H Spencer-Jones (UK)	
1958	R Peters (UK)	N Herlofson (Sweden)	E Herbays (Belgium)
1961	E F R Steacie (Canada) replaced Oct. '62 by S Horstadius (Sweden)	J van Mieghem (Belgium)	G R Laclavère (France)
1963	H W Thompson (UK)	D Blaskovic (Czechoslovakia)	G R Laclavère (France)
1966	J M Harrison (Canada)	K Chandrasekharan (India)	G R Laclavère (France)
1968	V A Ambartsumian (USSR)	K Chandrasekharan (India)	N B Cacciapuoti (Italy)
1970	V A Ambartsumian (USSR)	F A Stafleu (Netherlands)	N B Cacciapuoti (Italy)
1972	J Coulomb (France)	F A Stafleu (Netherlands)	N B Cacciapuoti (Italy)
1974	H Brown (USA)	J C Kendrew (UK)	D A Bekoe (Ghana)
1976	F B Straub (Hungary)	J C Kendrew (UK)	D A Bekoe (Ghana)
1978	C de Jager (Netherlands)	J C Kendrew (UK)	T F Malone (USA)
1980	D A Bekoe (Ghana)	J C I Dooge (Ireland)	T F Malone (USA)
1982	D A Bekoe (Ghana) replaced Nov. '83 by J C Kendrew (UK)	L Ernster (Sweden)	T F Malone (USA)
1984	J C Kendrew (UK)	L Ernster (Sweden)	K Thurau (Germany)
1986	J C Kendrew (UK)	L Ernster (Sweden)	K Thurau (Germany)
1988	M G K Menon (India)	J W M la Rivière (Netherlands)	K Thurau (Germany)
1990	M G K Menon (India)	J W M la Rivière (Netherlands)	M Petit (France)
1993	J C I Dooge (Ireland)	L J Cohen (UK)	M Petit (France)

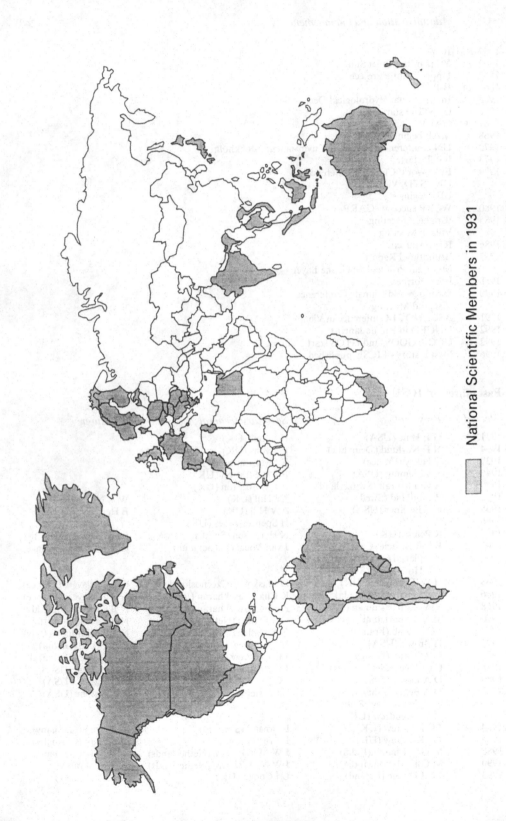

National Scientific Members in 1931

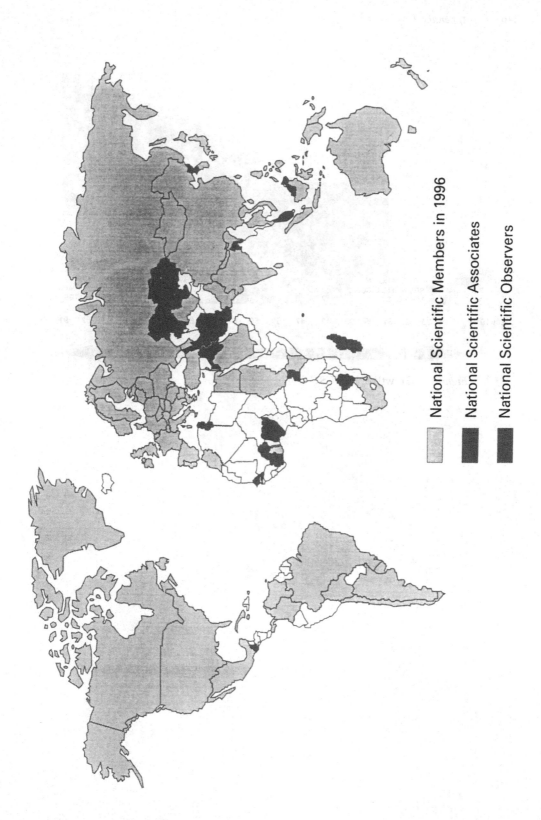

National Scientific Members in 1996

National Scientific Associates

National Scientific Observers

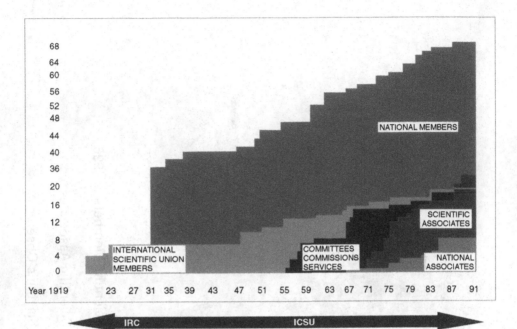

Figure 1 Growth of the ICSU family.

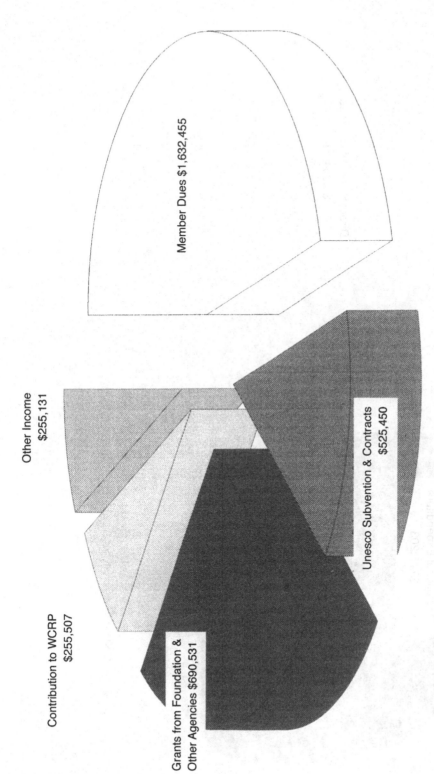

Member Dues $1,632,455

Other Income
$255,131

Unesco Subvention & Contracts
$525,450

Contribution to WCRP
$255,507

Grants from Foundation &
Other Agencies $690,531

Figure 2a   ICSU 1995 income distribution by source.

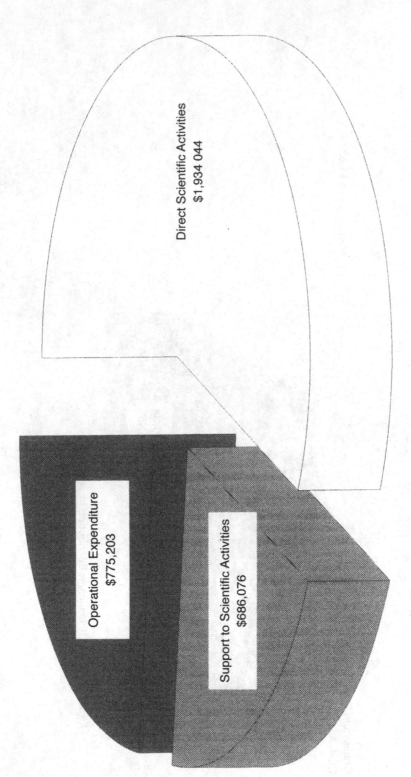

Direct Scientific Activities
$1,934 044

Operational Expenditure
$775,203

Support to Scientific Activities
$686,076

Figure 2b   ICSU 1995 expenditure distribution by type of activity.

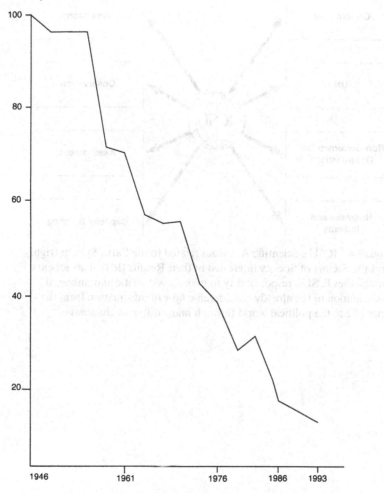

Figure 3   UNESCO Subvention as a percentage of the ICSU budget.

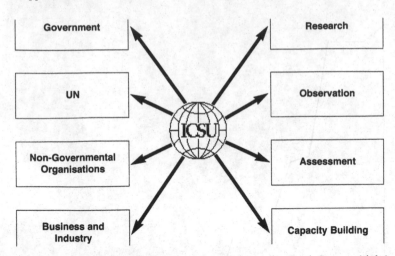

Figure 4   ICSU's Scientific Activities related to the Earth System (right) and the Sectors of Society interested in their Results (left). This schema emphasises ICSU's responsibility to make a systematic and unbiased contribution to the already considerable flow of information from the scientific to the political world through many different channels.

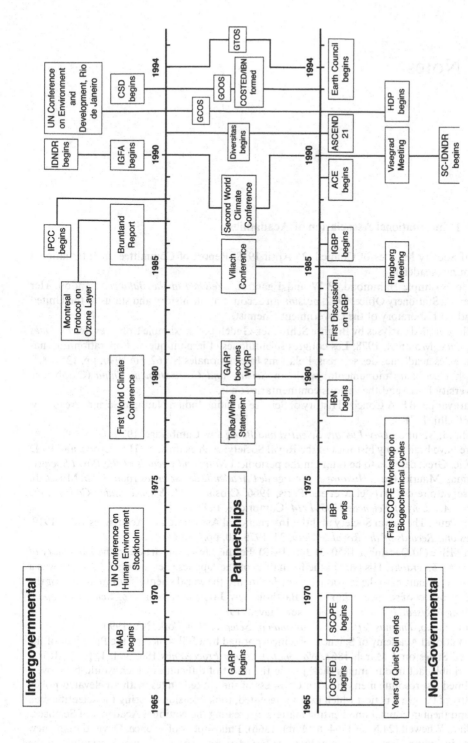

Figure 5 Developments in international science and policy-making in the global environment 1965–1994.

Notes

1 International Association of Acadamies

1 Royal Society Minutes of Council, 25 April 1918, Report of Committee on International Scientific Academies.

2 See, for example Hammond, P W and Egan, H, *Weighed in the Balance*. London, Her Majesty's Stationery Office, 1992, *passim*, an account of the history and status of the United Kingdom Laboratory of the Government Chemist.

3 See the detailed analyses by Brigitte Schroeder-Gudehus, for example from *Les Scientifiques et la Paix*, Montreal, 1978; Les congrès scientifiques et la politique de l'opération internationale des académies des sciences, Relations Internationales No 62, été 1990, pp. 135–48.

4 Joseph Needham's monumental work on *Science and Civilization in China* (Cambridge University Press) and the various commentaries on it.

5 Subarayappa, B S, A Concise History of Science in India, Indian National Science Academy, New Delhi, 1971.

6 Crosland, Maurice, *Gay-Lussac: Scientist and Bourgeois*, Cambridge, 1978.

7 There have been many histories of the Royal Society: such as those by H G Lyons, and by D McKie. Great detail is to be found in the periodical *Notes and Records of the Royal Society*.

8 Daumas, Maurice (ed.), *Histoire et Prestige de l'Académie des Sciences,1666–1966*. Musée du Conservatoire des Arts et Métiers, Paris, 1966. Crosland, M, *Science under Control: the French Academy of Sciences 1795–1914*, Cambridge, 1992.

9 Alter, Peter, The Royal Society and the International Association of Academies 1897–1919, *Notes and Records of the Royal Society*, 34, 1979–80, pp. 241–65

10 John Milne (30 December 1850–31 July 1913). For his life see the article in the *Dictionary of Scientific Biography*. His part in the formation of the Japanese Seismological Society was a major contribution to the introduction of Japan into the world scientific community. For the wider prospect of science in Japan see Bartholomew, J R, *The Formation of Science in Japan: Building a Research Tradition*, Yale, New Haven, 1989.

11 Bedini, Silvio, *Thomas Jefferson: Statesman of Science*, New York, NY, 1990.

12 The National Academy of Sciences was incorporated by a Bill signed by the President of the United States on 4 March 1863. *National Academy Proceedings* 1863 vol. 1, p. 7: ' It is a remarkable fact in our annals that, just in the midst of difficulties which would have overwhelmed less resolute men, the 37th Congress of the United States, with an elevated policy worthy of the great nation which they represented, took occasion to bring the scientific men around them in council on scientific matters by creating the National Academy of Sciences.'

13 William Whewell (24 May 1794–6 March 1866). Philosopher of science. Devised many new terms including electrical terminology for Faraday. First to use the word 'scientist' in the modern sense.

14 An account of the rise of chemical societies, national and international, will be found in W H Brock, *Fontana History of Chemistry*, London, 1992.

15 (Sir) Michael Foster (8 March 1836–28 January 1907). Cambridge physiologist. Biological Secretary of the Royal Society 1881–1903.

16 Price, Derek John de Solla, *Science since Babylon*, Yale University Press, New Haven, 1961, chapter 5.

17 There is no comprehensive history of scientific instrument industry. The Scientific Instrument Commission of IUHPS publishes periodical bibliographies.

18 Magnetische Verein, set up by Gauss and Weber. Results published in *Resultate des Beobachtungen aus den Magnetischen Vereins* (1833 to 1841).

19 Gauss, Karl Friedrich (30 April 1777–23 February 1855). See *Dictionary of Scientific Biography*.

20 Weber, Wilhelm Eduard (24 October 1804–23 June 1891). See *Dictionary of Scientific Biography*.

21 Bessel functions, developed by F W Bessel (1784–1864) from work of Euler, Lagrange, and others, for the study of planetary perturbations.

22 Helmert, Friedrich Robert (31 July 1843–15 June 1917) mathematics applied to gravitation and geodesy. Influential in establishment of International Bureau of Latitudes to monitor movement of the poles. See *Dictionary of Scientific Biography*.

23 Karlsruhe. A chemical conference convened there in 1860 to consider the composition of chemical substances. The conference was inconclusive, but Cannizzaro distributed a pamphlet which drew attention to a neglected hypothesis (1811) of Avogadro (1776–1856) which resolved anomalies in the determination of atomic weights. The whole question, central to chemical theory, was clarified from there on.

24 Etienne-Jules Marey (5 March 1830–15 May 1904). Pioneer of the use of graphical recording and cinematography in physiology. See entries in many biographical encyclopedias, and in histories of photography.

25 *Geography through a Century of International Congresses*, IGU/UNESCO, Paris 1972.

26 Schuster, A, *Nature*, 74, 12 July 1906, pp. 256–9.

27 His, Wilhelm, 9 July 1831– 1 May 1904, distinguished in anatomy and embryology.

28 Schuster, A, Nature, 74, 12 July 1906, pp. 256–9

29 Mommsen, Theodor (30 November 1817–1 November 1903) leading and influential classical historian.

30 Eduard Suess (20 August 1831–26 April 1914): Austrian geologist and academic administrator.

31 See Alter, P H, *Notes and Records of the Royal Society*, 34, 1979–80, pp. 241–265.

32 Royal Society Minutes of Council Meeting, 5 May 1898, p. 417

33 *International Catalogue of Scientific Literature*, 1901–1915. This grew out of the *Catalogue of Scientific Papers* published by the Royal Society of London, covering mainly the scientific literature of the 19th century. It was planned that an international catalogue should be published by the proposed international scientific body (eventually the IAA). On the outbreak of the 1914 war, the Royal Society withdrew and handed responsibility for various sections to individual scientific bodies.

34 Royal Society Minutes of Council, 1898, Minute 7, pp 430–1.

35 Royal Society Minutes of Council, 3 November 1898, Minute 11, p. 454.

36 See later chapters for IUPAC, IUPAP etc.

37 Royal Society Minutes of Council, 1899, pp. 84–9.

38 The nine other academies were: Amsterdam, Brussels, Budapest, Christiania, Copenhagen, Madrid, Stockholm, and two other French academies: Académie des Inscriptions et Belles Lettres, Académie des Sciences Morales et Politiques.

39 *The British Academy: The First Fifty Years*, Frederick Kenyon, London, 1952.
40 Jean-Gaston Darboux (14 August 1842–23 February 1917). Mathematician. Member of Académie des Sciences 1884. Secrétaire Perpétuel 1900–17. See the *Dictionary of Scientific Biography*.
41 The British survey of the 30th meridian or the triangulations associated with it would have needed to proceed along or near the shores of Lake Tanganyika, then in German territory. The procedure was to ask the Berlin Academy to approach its own government for the permissions necessary to admit surveyors or enable the Berlin Academy to initiate the necessary work.
42 See B Schroeder-Gudehus, *Les Scientifiques et la Paix*, Montreal, 1978.
43 For Fédération International de Documentation as co-operative activity with UNESCO see surveys of UNESCO work in librarianship and library organisation. For history of FID see *Encyclopedia of Library and Information Science*, edited by Allan Kent and Harold Lancour, Dekker, 1974, pp. 377–402. An English language version of the Universal Decimal Classification is published by the British Standards Institution as BS 1000. The preface to this Standard contains a history of the FID.
44 Prince Boris Galitzin (or Golitsin) (1862–1916) Russian geodesist, last President of International Association of Seismology (which lapsed in 1915). See entry under Golitsin in *Dictionary of Scientific Biography*.
45 See Wright, Helen, *Explorer of the Universe*, New York, NY, 1966.
46 Correspondence about Italian fears of being left out of discussion: Royal Society Minutes of Council, 14 February 1918.
47 Anglo-French discussions: Royal Society Minutes of Council, 1918–19 *passim* (see Alter, P *Notes and Records of the Royal Society*, 34, 1979–80, pp. 241–65.)
48 Schuster's views on future needs. See Alter, P, *Notes and Records of the Royal Society*, 34, 1979–80, pp. 241–65
49 *Royal Society Year Book,* 1919, pp. 172–3, 181–4.

2 The International Research Council (1919–1931)

1 See, for example, Crombie, A (ed.), Scientific Change, London 1963; Kuhn, T, *Structure of Scientific Revolutions*, Chicago, 1962,1970.
2 A typical account of a contemporary view of a centenary exhibition is in Ingram, J S, *The Centennial Exhibition . . . Commemorating the First Centenary of American Independence*, Philadelphia, 1876. This can be compared with the accounts of the Loan Collection of Scientific Instruments, London, 1876. The former is all popular enthusiasm, the latter constructive and academic. (See Greenaway, F, More than a mere gazing place, pp. 139–45 in: *Making Instruments Count*, eds: R G W, Anderson, Bennett, J A, Ryan W F., Variorum Press, Cambridge, 1993).
3 E S Ferguson, *Bibliography of the History of Technology*, Cambridge, Mass., 1968, pp. 192–200.
4 The detail of this kind of belief is to be found in studies of Auguste Comte (1798–1857) and his followers. For readers of this book a good starting-point is the entry under Comte in the *Dictionary of Scientific Biography* which has a good bibliography.
5 Biographies of William Crookes (1832–1919) and William Ramsay (1852–1916) tell the tale of Crookes' dramatic telegram (1895) to Ramsay: 'krypton is helium'; meaning that an element found on earth was identical with an element hitherto supposed to exist only in the sun.
6 There is a huge League of Nations literature. A few titles are: Murray, Gilbert, *From the*

League to the UN, London, 1952; Walters, F P, *A History of the League of Nations*, London, 1952; Northedge, F S, *The League of Nations*, Leicester, 1986.

7 Fritz Haber is a case in point. He was a discoverer whose work on the fixation of atmospheric nitrogen was to be of immense world-wide beneficial significance because of the use of nitrates in crop fertilising. But they were also invaluable as a source of explosives. In wartime, Haber felt devoted to a national military effort. His example is multiplied on both sides. See Coates, J E, 'Haber Memorial Lecture', in: *Memorial Lectures delivered before the Chemical Society 1933–1942*, vol. IV, London, 1951, pp. 127–57.

8 See the article on George Ellery Hale in the *Dictionary of Scientific Biography* which has a very extensive Hale bibliography. A useful introduction is Helen Wright, *Explorer of the Universe*, New York, 1966. There is good deal on Hale's character, and his ability to get on good terms with colleagues. The entry in *Obituary Notices of Fellows of the Royal Society*, (2, (1936–8), pp. 523–9) deals only with his scientific work.

9 See the *Dictionary of Scientific Biography* for Schuster's scientific work, Royal Society obituary for opinions of colleagues. (*Obituary Notices of Fellows of the Royal Society*, I (1932–5), pp. 409–23.) Schuster's autobiographical *Biographical Fragments* London 1932, has little bearing on his international interests. These were expressed in a Royal Institution lecture (18 May 1906): International Science; Nature, 74, 1906, p. 205.

10 See for example R W Home (ed.), *Australian Science in the Making*, Cambridge, 1989, for a survey of a Dominion's independent expansion into many scientific fields.

11 See Note (6).

12 See tables of successive membership of IRC, ICSU, in Appendix 2. See also *Royal Society Year Book*, 1920 (entry for 26 February) p. 470.

13 Gravis, Jean-Joseph Auguste (29 August 1857–2 January 1937). Vice-Directeur (and later Directeur) de la Classe des Sciences de l'Académie Royale de Belgique

14 Helen Wright's biography of G E Hale (chapter 1, note 39) gives a good account of personalities involved in the formation of the NRC.

15 The title *United Kingdom* is often used in a loose sense, the scope of its relevance having been changed from time to time. The most important change for the purpose of this history was that following the separation of the southern counties of Ireland to form the Irish Free State (Eire) (1923).

16 Serbia became the leading element in the new state of Yugoslavia after the settlement of boundary lines by successive peace negotiations (notably the treaty of Versailles), and the incorporation of other territories. At the time of writing this note the status of the whole region is uncertain.

17 Discussion of Statutes of 1919 is to be found in Schroeder-Gudehus (see chapter 1, note 42), pp. 116–22.

18 In the archives of the Royal Society there is a letter from Picard to Schuster which breaks off in the middle as Picard suddenly tells Schuster of the death of his second son as the delayed result of the First World War, a war which had taken his whole family from him. But Picard continues with his obligation to make a success of IRC.(1925)

19 Fennell, Roger, *History of IUPAC*, Oxford, 1994.

20 The International Geodesy Association originated as the Mitteleuropäische Gradmessung (1862), an association mainly of German-speaking states, which extended its membership in 1867 to become a European Geodetic Association. At its 1883 conference, the USA and Great Britain were represented, and in 1886 the name of International Geodetic Association was adopted. After 1918 the Association needed to be reconstituted, which came about with the formation of the IRC. The association was then superseded by the International Union of Geodesy and Geophysics.

21 There had been International Mathematical Congresses since 1897. Draft statutes of an

International Mathematical Union were discussed during the Constitutive Assembly of IRC in 1919. An interim organisation was set up. IMU came into existence on 20 September 1920 in Strasbourg, with de la Vallée Poussin as President. Picard and Volterra were made Honorary Vice-Presidents so making a strong link with IRC. It did not withdraw from IRC; its membership of the successor ICSU expired when IMU was dissolved in 1932. A new constitutive meeting was held in 1950; IMU was formally recreated in September 1951, and readmitted to ICSU in 1952.

22 Translation of Picard, p. 212:

> Like any enterprise of a new kind it [IRC] has met, here and there, some scepticism. To unite, be it by a rather light tie, so many Associations, has seemed chimeric to some who prefer absolute independence for various international scientific organizations. Short though our history may be, it seems to show that these fears are without foundation. The authority of the International Research Council over these different bodies is quite light. Moreover: one can believe that certain Unions have been happy to leave to the International Council the trouble of taking some decisions which they are not anxious to take themselves.

23 Picard 1925: 'The events of which the world has been the theatre a few years ago, have sternly reminded us of certain truths too often forgotten. You will see one day on what conditions and to what extent it will be desirable to cast a veil over the past.'

24   There are areas in which collaboration with the central powers is essential: indeed it has already begun, although it may be through complex and indirect paths. There exist, outside IRC, organisations in which this collaboration is regularly utilised, and several commissions set up by the League of Nations are an example.

    We believe that the moment has come to bestow on scientific efforts the character of universality which they should possess as fully as possible, because of the nature of science itself, and we consider that one can take this step without hesitation.

Lorentz probably had in mind the Institute for International Intellectual Co-operation which had been formed in 1922 as a result of a joint British–French initiative, following the discussions of the League of Nations Committee on Intellectual Cooperation (first meeting 1 August 1922). See chapter 3 below, *passim*.

25 Accounts of the Locarno 'pact' are to be found in major political histories of the period. It was in essence a renunciation of resort to war, and created an atmosphere of optimism. One immediate consequence was the entry of Germany into the League of Nations. The detail of the diplomatic history, of which the Locarno pact is part, is beyond our scope.

26 H Spencer-Jones, The early history of ICSU, *ICSU Review*, 2(4), 1960, p. 179.

27 Schuster: see note 9 above.

3 ICSU: beginning and establishment

1 G M Trevelyan, *passim* in *Clio: a Muse*, London, 1913.

2 The new name emerged from discussion in the Commission appointed in 1928.

3 The full text of Picard's speech is to be found in the *Reports of Proceedings of the Fifth General Assembly of the International Research Council and of the First General Assembly of the International Council of Scientific Unions*, Brussels, 1931.

4 W de Sitter (6 May 1872–20 November 1934) in H Spencer-Jones, The Early History of ICSU, *ICSU Review*, 2, 1960, p. 176.

5 Marconi, Guglielmo (25 April 1874–20 July 1937) had become part of the British scientific

establishment after his considerable commercial success. Oliver Heaviside (1850–1925) had proposed what became known as the ionosphere. A E Kennelly (1861–1939), who had made similar proposals, was a United States delegate to the 1931 General Assembly.

6 The International Union of Geological Sciences was formed as an additional resource to an already well-established international system of quadrennial congresses. IUGS was formed to bring its members into closer liaison with other scientists active on the international scene.

7 Vote in Geographical Union on a population commission. See report of IUG in 1934 General Assembly report.

8 P P Ewald (ed.), *Fifty years of X-ray Diffraction*, Utrecht (for IUCr), 1962.

9 Volterra, Vito (3 May 1860–11 October 1940). Mathematician. Active in politics. Opposed Fascism. Refused to sign oath of allegiance to Mussolini 1932. Deprived of all positions and honours. Elected to Pontifical Academy of Sciences 1936. Travelled widely. Died in Rome. See *Dictionary of Scientific Biography, Enciclopedia Italiana.*

10 Ferrié, Gustave-Auguste (19 November 1868–16 February 1932). Pioneer in wireless telegraphy. Head of wireless service in France in 1932. Member of Académie des Sciences 1922.

11 Nörlund, Neils Erik (1885–1981). Danish mathematician. Director of Danish Geodetic Institute 1922–55. President of Danish Royal Academy of Sciences (1927) and of ICSU (1934–7). See *Biographical Memoirs of Fellows of the Royal Society*, XXIX, 1983, pp. 481–93.

12 Pelseneer, Paul (26 March 1863–5 May 1945). Belgian biologist. Permanent Secretary of the Belgian Royal Academy of Sciences. Member of the French Academy of Sciences.

13 Went, Friedrich August Ferdinand Christian (18 June 1863–24 July 1935). Botanist. President Royal Netherlands Academy of Sciences.

14 Sir Henry Lyons, celebrated as Director of the Science Museum, South Kensington (for which see David Follett, *The rise of the Science Museum under Henry Lyons*, London, 1978. This work makes no mention of Lyons' connection with ICSU, for which see *Biographical Memoirs of the Royal Society*, IV, 1944, pp. 795–805.

15 See *Reports of Proceedings of the Fifth Assembly of the International Research Council and of the First General Assembly of the International Council of Scientific Unions*, H Lyons ed., 1931, French version p. 23, English version p. 78.

16 *UNESCO document UIS-89/WS/5 October 1989* lists over 100 publications of IIIC of which only 4 are scientific. Three are reports of conferences organised in collaboration with Unions.

4 ICSU development 1931–1939

1 International Postal Union, originated in United States proposals, leading to adoption in Berne in 1875 of a International Postal Convention, which governs the operation of the active body, the International Postal Union.

2 The Leni Riefenstal film of the 1936 Olympics is a striking demonstration of the political use of new communication techniques.

3 For the history and development of ICOM the best source is the UNESCO periodical *MUSEUM.*

4 For the techniques of telecommunications up to about 1950, see E G Tucker, chapter 50, in T I Williams (ed,), *History of Technology* VII, Oxford, 1978.

5 See Chapter 1, note 10.

6 Baker, F W G, First International Polar Year 1882–83, *Polar Record*, 21(132), September 1982; *Commémoration des Années Polaires et de l'Année Géophysique*, Palais de la Découverte, Paris,1984.

7 The complexity of the problem is to be seen in a comparison of two books on the same theme: Landes, David, *Prometheus Unbound,* Cambridge, Mass., 1969, and Salomon, Jean-Jacques *Prométhée empêtré*, Paris 1981.

8 Mees, C E K and Leermakers, J A, *The Organisation of Scientific Industrial Research*, New York, 1950. It is instructive to compare this 1950 study with the book of the same title by Mees published in 1920. The earlier work sees industrial research as akin to academic research. The later work, written when Mees had a lifetime of experience behind him, makes it appear quite different in form and function.

9 Jewkes, John et al.*The Sources of Invention*, New York, 1962.

10 There is a large literature of the scientific discoveries of the 1930s, emphasising particularly those in atomic physics. See for example: *Scientific Thought 1900–1960*, ed. Harré, R, Oxford 1969; Unesco, *History of Science*, vol. 5.

11 See, for example, ICSU, *Report of Second General Assembly* (Brussels) London, 1935, p. 19.

12 For the invention and development of the quartz oscillator see the life of George Washington Pierce (11 January 1872–25 August 1956) (Saunders, F A and Hunt, F V, *Biographical Memoirs, National Academy of Sciences*, vol. 33, 1959, pp. 351–80). For an interesting indication of the way the quartz oscillator was about to supersede the mechanical and electrical devices used in international time-keeping see F Hope-Jones, *Electric Time-keeping*, London (first edition 1940, second edition 1949, facsimile 1976) chapter 27.

13 An example of the bad effect of separation is that of the emergence during the war of two conflicting theories of stellar atmospheres (Unsöld and Menzel). By the time they could be resolved both schools of thought were becoming obsolete.

14 Berkeley Students' Observatory was inaugurated on 30 January 1904 (*Publications of the Astronomical Society of the Pacific*, 16(95) and was still in operation in 1946. Short comment in Oesterbrock, D E, Armin O Leschner and the Berkeley Astronomical Department. *Astronomy Quarterly*, 7, 1990, pp. 95–115.

15 Johann Jacob Baeyer (5 November 1794–11 September 1885). Lieutenant-General, father of the celebrated chemist, Adolf von Baeyer (1835–1917). J J Baeyer worked first with the astronomer Bessel on official Prussian latitude and longitude measurements, and then initiated a European programme.

16 Fennell, Roger *History of IUPAC*, Oxford 1994.

17 It is not clear from condensed reports what they were talking about. Direct photography of documents was practised for archival and legal purposes. There were ponderous processes like Photostat. There were dyeline processes much used for copying technical drawings. None of these was of the slightest use economically for dissemination of publications. Microfilming was just being developed, but for record purposes only.

18 Bakelite: see Williams, T W, *History of Technology*, VII, Oxford, 1978, p. 555; Mossman, S J, Morris, P J T (eds.), *The Development of Plastics*, (Royal Society of Chemistry) London 1994.

19 For life of Carothers, Wallace Hume, see: Adams, R, *Biographical Memoirs, National Academy of Sciences*, 20, 1939, pp. 293–309. For description of research and development of nylon see Hounshell, D A, and Smith, J K, *Science and Corporate Strategy: DuPont R & D 1902–1980*, Cambridge University Press, Cambridge Mass., 1988.

20 Histories of penicillin assess different degrees of credit for achievement. Biographies of Fleming and Florey, each with substantial bibliographies, are in *Biographical Memoirs of Fellows of the Royal Society*.

21 Ewald, P P et. al., *Fifty Years of X-ray Diffraction*, (for International Union of Crystallography), Utrecht, 1962.

22 de Martonne, Emmanuel-Louis-Eugène (1 April 1873–24 July 1955). Head of Department of Geography at the Sorbonne 1909–44. Organised meetings of IGU from 1931 to 1938,

President of IGU 1949 to 1955. Author of valuable reference works. (See *Dictionary of Scientific Biography* for detail and for bibliography.)

23 For history of exploration see *History of World Exploration*, Keay J, (ed. for Royal Geographical Society) London, 1991.

24 Humbert, Jean Henri,(24 January 1887–20 October 1967). Botanist, Professor at Museum of Natural History, Paris.

25 What may be the last surviving ultra-centrifuge of that generation, made under Svedberg's supervision, survives in the Science Museum, South Kensington, London.

26 Some British examples are: Arabella Buckley, *A Short History of Natural Science*, London, 1888; Robert Routledge, *Discoveries and Inventions of the Nineteenth Century*, London, 1901.

27 See Moran, J, Printing chapter 51 in: Williams, T I (ed.), *History of Technology*, VII, Oxford, 1978.

28 See, for example, Price, Derek John de Solla, *Little Science, Big Science*, Columbia University Press, New York, 1962.

29 A useful survey of technical progress in the first half of the 20th century is to be found in volumes 6 and 7 of Williams, T I (ed.), *History of Technology*, Oxford, 1978. A wider survey of aviation with detail of commercial applications is in Gibbs-Smith, C H, A*viation: an Historical Survey*, London, 1970. A vivid pictorial history with useful chronology is Gibbs-Smith, C H, *Flight Through the Ages*, New York, 1974.

5 ICSU in wartime

1 This view is exemplified in the title first given to one of the Smithsonian Institution Museums: National Museum of History and Technology. This was later changed to National Museum of American History, perhaps a recognition of the fact that technology should be studied as an essential component of historical change.

2 PLUTO (the pipe-line under the ocean) is an example of a technological achievement which might have been conceived for peaceful use. The intensive German search for conversion of coal to petroleum is equally a line of research not necessarily military in significance. Both figure in peace-time research and application.

3 Jones, R V, *Most Secret War*, London, 1978, deals mainly with radio detection and clandestine communication, but also reveals a great deal about other activities which had a scientific core.

4 The history of the synthesis and use of DDT is important in considering the control of insect vectors. It also carries over into the later concern for ecological problems.

5 Home, R W, in R W Home (ed.), *Australian Science in the Making*, Cambridge, 1988, pp. 220–51.

6 J M Burgers (1895–)

7 La Cour, Dan. (1876–1942) Born Copenhagen. President of IUGG 1936–42. General Secretary of the International Association of Terrestrial Magnetism and Atmospheric Electricity, 1933–6.

8 On the work of French scientists in atomic physics from 1930 to 1948 see Weart, Stephen, Science and Power, Harvard, 1979.

9 For the history of IAU until 1970 see Blaauw, Adriaan, History of IAU: *The Birth and First Half Century of the International Astronomical Union*, Kluwer, Dordrecht, 1984.

10 Fennell, Roger, *History of IUPAC: 1919–1987*, Oxford, 1994.

11 Taylor, H S (1890–1974). See *Biographical Memoirs of Fellows of the Royal Society*, XXI, 1975, pp. 517–47.

12 Timmermans, J E L (1882–1971).

13 Louis de Broglie. See *Dictionary of Scientific Biography*.
14 (Sir) Edward Victor Appleton, see entries in *Dictionary of Scientific Biography, Biographical Memoirs of Fellows of the Royal Society*, XII, 1966, pp. 1–21. President of URSI (1934–52). Chairman of Mixed Commission on the Ionosphere (1947–60). Was one of those who initiated the International Geophysical Year of 1957–8.
15 Abraham, Henri (1868–1943). Secretary-General IUPAP 1922–43. Died in a concentration camp.
16 See Kamminga, Harmke, The International Union of Crystallography: its formation and early development, *Acta Crystallographica* (1989) A45, pp. 581–601.
17 See *Geography through a Century of International Congresses*, UNESCO, 1972.
18 Colin, Elicio. Directeur de Bibliographie Géographique Annuelle. Association de Géographes Français.
19 Hartmann, Max. Director of Max Planck Institute of Biology at Tübingen 1944. Kühn, Alfred (1885–1968). Second Director of Kaiser Wilhelm Institute in Berlin-Dahlem, 1937–45. Director of Max Planck Institute Tübingen 1945–64. Wettstein, Fritz (24 June 1895–12 February 1945.)
20 *IUGG. Account of the War Years 1939–45*. ed. Stagg, J, Cambridge, 1946.

6 A new beginning (1945)

1 For Atlantic Charter see any of the major biographies of the main wartime leaders.
2 For the general history of the UN see Luard, Evan, *A History of the United Nations*, London, 1982–9, volumes 1 & 2; Brooms, Bengt, *The United Nations*, Helsinki,1990. For the text of the Dumbarton Oaks meeting, see 'Dumbarton Oaks Conversations on World Organization,' in: *United Nations Documents (1941–1945)*, Royal Institute of International Affairs, London, 1946.
3 Charter of the United Nations and Statute of the International Court of Justice, Department of Public Information, United Nations N. Y. 10017 (DPI/511), 87 pp.
4 The part played by the Conference of Allied Ministers of Education in the establishment of UNESCO is described in Wells, Clare, *The UN, UNESCO, and the Politics of Knowledge*, London, 1987. For CAME see pp. 116–17. Documentary sources are to be found in UNESCO Archives under: *AG2 Conference of Allied Ministers of Education (CAME) 1942–1945 London*.
5 Sources for the history of UNESCO are listed in *Unesco Archive Finding Aids, ARC.91/WS/2*. The text of the UNESCO constitution can be found in the *United Nations Documents (RIIA 1946)*. See also Laves, W and Thomson, Charles, *Unesco, Purpose, Progress, and Prospects*, Indiana University Press, Bloomington, 1957.
6 Sir Alfred Zimmern. (26 January 1879–24 November 1957). Internationalist. Deputy Director IIIC 1926–30. See entry in *Dictionary of National Biography*.
7 (Sir) Julian Sorell Huxley (22 June 1887–14 February 1975). See *Biographical Memoirs of Fellows of the Royal Society*, XXII, 1976.
8 Needham, Joseph, *Memorandum on place of science and international scientific cooperation in post-war world organization*, 15 March 1945. An appendix to this memorandum is a very full survey of the discussion around the foundation of UNESCO.
9 Pierre Auger (14 May 1899–24 December 1993). See obituary notice by Alain Gille, *Lien (Newsletter of the Association of Former Unesco Staff Members)*, 47, pp. 43–4.
10 (Sir) Robert Robinson (13 September 1886–8 February 1975). Nobel Laureate. See *Biographical Memoirs of Fellows of the Royal Society*, 1976; *Dictionary of National Biography*, Williams, T I, *Robert Robinson, Chemist Extraordinary*, London, 1990.
11 H R Kruyt, President of Division of Sciences, Netherlands Academy of Sciences.
12 Fabry, Charles (1867–1945). French physicist. First Director of the Institut d' Optique (Paris). Elected President of ICSU but soon resigned and died shortly afterwards.

13 Eddington (Sir) Arthur (1882–1944). Mathematician and astrophysicist. President IAU (1938–44). *Obituary Notices Fellows of the Royal Society*, V, 1945, pp. 113–24.

14 H Abraham, see chapter 5, note 15.

15 First draft of agreement appears in *Report of General Assembly 1946*. Subsequent revisions appear regularly in ICSU publications, for example at present in ICSU *Year Book*.

16 J J Mayoux, Director IIIC, later Staff Member for Cultural Affairs at UNESCO.

17 Angel Establier (born 1905). Biochemist. Assistant Director of the Science Section of IIIC of the League of Nations) from 1931 to 1941, and from 1945 until the dissolution of IIIC in 1946.

18 Text of letter of 'denunciation':

> Paris le 16 juin 1946
> Monsieur le Président,
>
> Nous avons l'honneur de vous informer que, conformément à l'article 7 de l'Accord conclu entre le Conseil international des Unions scientifiques et l'Organisation internationale de Coopération intellectuelle, l'Institut international de Coopération intellectuelle agissant en qualité d'organe exécutif de l'Organisation internationale de Coopération intellectuelle vous notifie la dénonciation du dit accord.
>
> Nous vous serions obligés de bien vouloir nous donner acte de cette dénonciation et nous confirmer votre accord.
>
> Veuillez agréer, Monsieur le Président, l'assurance de ma haute considération.
>
> J J Mayoux (Le Directeur)

19 Hadamard, Jacques Salomon (8 December 1865–17 October 1963). French mathematician. Professor at Sorbonne, Collège de France, Ecole Polytechnique. Member Académie des Sciences, Fellow Royal Society. At Princeton during WWII. (*Biographical Memoirs of Fellows of Royal Society*, 1965, pp. 75–100.)

20 A V Hill (26 September 1886–3 June 1977) *Biographical memoirs of Fellows of Royal Society*, 24 pp. 71–150.

21 ICSU: *Report of Proceedings of Fourth General Assembly, London 1946*, Cambridge, 1946.

7 ICSU in a post-war world

1 *Science: the Endless Frontier, A Report to the President [of the United States] on a Program for Postwar Scientific Research*. Vannevar Bush, July 1945. (Reprinted July 1960) National Science Foundation, Washington DC.

2 Morris, P J T, *The American Synthetic Rubber Research Programme*, Philadelphia, Pennsylvania University Press, 1989.

3 MacFarlane, Gwyn, *Howard Florey: the Making of a Great Scientist*, Oxford 1979; Williams,Tevor I, *Howard Florey: Penicillin and After*, Oxford 1987.

4 History of television. Accounts of the main processes and inventions involved in television will be found in the McGraw Hill *Encyclopedia of Science and Technology*. An extensive bibliography listing works on both technical and social advance is appended to Sarlemlin, A, and De Vries, M, The piecemeal rationality of application-oriented research, pp. 99–132 in: Kroes P and Bakker M, *Technological Development and Science in the Industrial Age*, (*Boston Studies in the Philosophy of Science*), Kluwer, Dordrecht/Boston/ London, 1992.

5 See Rosen, E S, The development and characterization of the intraocular lens. chapter 5, p. 50 in: *Intraocular Lens Implantation*, eds., E S Rosen, W M Haining, and E J Arnott, 1984, St Louis.

6 The events are described in personal terms in Watson, James D, *The Double Helix*, London, 1968.
7 Carothers: see chapter 4, note 19.
8 See previous note on Establier, chapter 6, note 17.
9 Ronald Fraser, New Zealander: formerly on UNESCO staff.
10 National representatives. See Appendix 2.
11 Raman See *Biographical Memoirs of Fellows of the Royal Society*, XVII, 1971, pp. 565–92.
12 Needham, Joseph, *Science and Civilization in China*, work in progress, many volumes since 1954 on Chinese science.
13 Pacific Science Association: see chapter 9.
14 See Baker, F W G, *ICSU-UNESCO; Forty years of Co-operation*, ICSU, 1986.
15 SCOR: see chapter 9.
16 Intergovernmental Oceanic Commission: see chapter 18.
17 COSPAR: see chapter 9.
18 CETEX: for origin see *ICSU Review*, 4, 1959 pp. 100–5.
19 Crawford, Elisabeth, The Universe of International Science, in: *Solomon's House Revisited*, (ed) Frängsmyr, T, Nobel Symposium, 75, 1990, pp. 251–69.

8 The free conduct of science

1 UN *Declaration of Human Rights* adopted 10 December 1948 by UN General Asembly.
2 G E Hale: retiring Presidential Address 1934: *Report of Second General Assembly of ICSU*, 1934, pp. 4–7.
3 R M Cooke, (ed.), *Refugee Scholars: Conversations with Tessa Simpson*, London, 1992.
4 Eighth General Assembly 1958. Reported in ICSU Review, 1, pp. 144–5.
5 North Atlantic Treaty Organisation: formed 1949, original members were United States, Canada, United Kingdom, France, Belgium, Denmark, Iceland, Italy, Norway, Portugal, The Netherlands. Greece and Turkey joined in 1952, and the Federal Republic of Germany in 1955.
6 *1974 Recommendation on the Status of Scientific Researchers. Eighteenth Session General Conference of UNESCO*, Paris, 17–23 November 1974.
7 SCFCS as 'watchdog': *Science International*, September 1991 (60 Years issue), pp. 31–2.
8 First published as *Advice to Organisers of International Scientific Meetings*. Replaced by *Universality of Science: Handbook of ICSU's Standing Committee on the Free Circulation of Scientists*, 1989.
9 Canada refused visas to two USSR scientist for entry to attend the 1984 General Assembly in Ottawa. Censure at the General Assembly followed. Canada later expressed regret and gave an assurance that the event was an isolated incident and that freedom of entry would be observed in the future.
10 Personal communication.
11 For detail of this discussion see official record of debate at 22nd General Assembly, Beijing 1988 (ICSU, Paris).
12 Repudiation: a requirement that an applicant for a visa declare that he is not in agreement with any law of his own country restricting free movement.
13 *Universality of Science*, section V, paragraph 17.

9 Living machinery: officers and staff

1 See the discussion of relations with UNESCO in the Report of the 1946 Fourth General Assembly (1946), pp. 49–55.

2 Huxley, Julian, *Memories II*, London, 1973, pp. 7–21.

3 Bhabha, Homi Jehangir (30 October 1909–24 January 1966). See entries in *Dictionary of Scientific Biography*, (supplementary volume). *Biographical Memoirs of the Royal Society*, XIII, 1967 pp. 35–52.

4 A Establier. See chapter 6, note 17.

5 Ronald Fraser. See chapter 7, note 9.

6 A V Hill: chapter 6, note 20.

7 The successive homes of the Royal Society are described in *The Royal Society at Carlton House Terrace. London*, Royal Society, London, 1967.

8 Reports of General Committee, Laxenburg September 1975, General Assembly, Washington DC 1976.

9 History of rocketry: see *Cambridge Encyclopedia of Space. Space Technology*, ed. K Gatland.

10 Nobel Prize for Medicine or Physiology 1962 awarded to F H C Crick, J D Watson, and M H F Wilkins *'for their discoveries concerning the molecular structure of nucleic acids and its significance for information transfer in living material'*.

10 Growth: within science and outwards

1 Price, Derek John de Solla, *Little Science, Big Science*, Columbia University Press, New York, 1959.

2 For statistics on numbers of scientists in most countries see *Unesco World Science Report 1993*.

3 Kamminga, H. See chapter 5, note 16.

4 See Fennel, Roger, *History of IUPAC*, Oxford,1994.

5 IUHS, founded in 1947 following mainly French initiatives. It was created by the International Academy of the History of Science which had been founded in 1928. There was a succession of congresses from 1929 (as IAHS) followed by a joint IAHS/IUHS congress in 1948.

6 Boyd-Orr, Lord (John) 1880–1971. Nobel Peace Prize 1949. *Biographical Memoirs Fellows of Royal Society*.

7 IEEE (the USA body of that title).

8 History of IUMS by Kupferberg, E D. See 'The International Union of Microbiological Societies: A Brief History.' *American Society for Microbiology News*, 59, 1993, pp. 69–75. A full length manuscript version of this history is deposited in the IUMS/ASM archives at the Center for the History of Microbiology in the Kuhn Library, University of Maryland, Baltimore County.

9 Jones, Graham, *Role of Science and Technology in Developing Countries*, Oxford, 1971. Note particularly the work of P M S Blackett.

10 For WDCs see chapter 12.

11 ICSU at mid century

1 Stratton, F J M, 1881–1960. Astronomer and distinguished teacher of astronomers at Cambridge. Active in administration of many important bodies including British Association for the Advancement of Science and International Union of Astronomy. Secretary-General of ICSU from 1937 to 1952.

2 Hill, A V see chapter 6, note 20.

3 Noyes, W A (1857–1941). Chemist, noted for very precise analytical determinations at US Bureau of Standards and at Illinois. First Editor of *Chemical Abstracts*.

4 Laclavère, Georges. (1905–94). Geodesist, lengthy service with IUGG and notable Treasurer of ICSU 1961–68. Obituary in *Science International*, 58, December 1994, p. 47.

5 History of URSI. See: *URSI Golden Jubilee Memorial*, published by URSI from UCCLE Brussels 1968. Articles by E V Appleton, J H Dellinger, W J G Beynon. See also for period 1945–79: URSI after World War II, W J G Beynon (Sir Granville Beynon), in: *Proceedings of 60th Anniversary Colloquium of URSI*, Brussels, September 17–18, 1979.

6 Fennell, Roger, *History of IUPAC*, Oxford, 1994.

7 Personal experience.

8 This reference to a 'Cold War' oversimplifies the political–military issues which would need a discussion of relations between, for example, NATO, the Warsaw Pact Countries, the USA and China, the USSR and China. These are beyond our scope here.

9 The membership of the Korean People's Republic lapsed after a while. At the time of publication its outcome remains uncertain.

10 Professional organisation: see, for example, Russell, C A, Coley, N G, and Roberts, G K, *Chemists by Profession*, Open University Press, Milton Keynes, 1977.

11 Hogben, L, *Science for the Citizen* London 1938.

12 ICSU *Report of Third General Assembly*, London, 1937.

12 The International Geophysical Year and its heritage

1 For initiation of FAGS see Report of 8th ICSU General Assembly, *ICSU Review*, 1959, pp. 57–101, sections on FAGS: pp. 75–6, 81–3.

2 Laclavère, G., see chapter 11, see note 4.

3 The detailed history of the antecedents of IGY and the preceding IPYs is to be found in the *Annals of IGY*, vol. I London, 1959.

4 International Meteorological Commission. See *Annals of IGY*, vol. I.

5 van Allen, James Alfred (1914–). Physicist. Discoverer of radiation belts forming earth's magnetosphere.

6 Sydney Chapman (29 January 1888–16 June 1970) See *Dictionary of Scientific Biography, Biographical Memoirs of Fellows of the Royal Society*.

7 Marcel Nicolet (26 February 1912–). An interview in WMO Bulletin, 39(4), 1990, pp. 235–46, gives a detailed account of Nicolet's career and place in the history of IGY.

8 Appleton, see chapter 5, note 14.

9 See *Annals of IGY*, vol. II.

10 Sixth ICSU General Assembly, Amsterdam, 1952.

11 Text of resolution. *Annals of IGY*, vol. I, p. 384.

12 Herbays, E. Secretary-General URSI, 1963.

13 Finances of IGY. *Annals of IGY*.

14 Pergamon Press was one of the earliest ventures of Robert Maxwell, whose financial career and mysterious death belong to a more sensational history than this.

15 Batisse, M, *Intergovernmental Cooperation, in World Science Report 1993*, UNESCO, 1993, pp. 152–66.

13 Data and scientific information

1 For WDCs see Chapman, Sydney, *ICSU Review*, vol. 1, January 1959, pp. 16–26.

2 Rossini in CODATA Newsletter 1, *First CODATA CONFERENCE 1968*.

3 Price, Derek John de Solla, *Little Science, Big Science*, Columbia University Press, New York, 1959.

4 See Report of 8th General Assembly of ICSU, 1958, *ICSU Review*, 1, 1959, pp. 57–101. Section on FAGS, pp. 75–6, 81–3.
5 CODATA Newsletter. 62, February 1993, pp. 2–3.
6 *Acta Crystallographica*: foundation. See chapter 5.
7 Harrison Brown: President ICSU 1974–76
8 CODATA Newsletter 58, November 1991 (issue devoted to anniversary).
9 For members of original CODATA team see CODATA Newsletter 58, November 1991.
10 *ICSTI: Review of Present and Past Activities*, ICSTI, Paris, February 1993.
11 Boutry, G-A, The ICSU Abstracting Board, *ICSU Review*, 1, 1959 pp. 113–37.
12 *Science Citation Index*: annual publication giving information about how many times a work is referred to in other works. Covers period from 1961.
13 Auger, Pierre (16 May 1899–24 December 1993). Obituary by Alain Gille in *Link Bulletin* (Newsletter of Association of Former Staff Members of UNESCO), No 47, 1944, pp. 43–4.
14 *Chemical Abstracts*. An abstracting facility (1895) in an MIT publication (Review of American Chemical Research), edited by A A Noyes (1866–1936) was taken over by W A Noyes (1857–1941), editor of the *Journal of the American Chemical Society*, who was inspired by it to establish a substantial independent abstracting publication under the title Chemical Abstracts. See Brock, W H, *Fontana History of Chemistry*, London, 1992, p. 452.
15 Royal Society definition of a good synopsis in *Guide for the Preparation and Publication of Synopses* (1949). See Boutry reference in note 11 above, p. 127.
16 Boutry, *ibid.*, p. 127.
17 This withdrawal from ICSU AB was unrelated to the withdrawal of UK and USA from UNESCO.

14 World projects and the environment

1 Rudolph Peters (Sir) (1909–82).
2 Waddington, C H, in *The Evolution of IBP*, ed. Worthington, E B, Cambridge, 1975, p. 5.
3 *Evolution of IBP* (ed. Worthington) p. 6.
4 IUCN: International Union for the Conservation of Nature and Natural Resources founded 1948.
5 Smith, F E, Speech to Plenary Session of National Research Council (*Proceedings of the National Academy of Science*, 60, 1968, pp. 1–50).
6 Worthington, E Barton, *The Ecological Century, A Personal Appraisal*, Oxford, 1983.
7 Ronald Keay in *The Evolution of IBP*, ed.Worthington, E B, p. 129.
8 Most of the original material produced during IBP is deposited with the Linnaean Society of London (Worthington, *Ecological Century*, p. 166). Publication of IBP work was mainly through existing journals in the several disciplines involved.
9 This account is indebted to White, G F, SCOPE, the First Sixteen Years. *Environmental Conservation*, 14(1), Spring 1987, pp. 7–13.
10 IRPTC. There were 128 National Correspondents from 120 countries at October 1994. IRPTC then held 88,000 records on over 1,000 individual compounds.
11 ENUWAR publications; *SCOPE 28. Environmental Consequences of Nuclear War*, volumes I and II,
12 SCOPE Report on Villach meeting. SCOPE 29. *The Greenhouse Effect: Climate Change and Ecosystems*. Report (533 pp.) presented to the Villach Conference. The Conference Declaration is to be found on pp. xx–xxiv.

15 ICSU and UNESCO

1 For a simplified account of UN, origin, present organisation, and associated bodies see *Basic Facts About the United Nations*, Department of Public Information, United Nations, New York, NY, 1993 (ISBN 92 1 100499 3) or later edition.
2 Reference to UNESCO decision. Baker, F W G, *ICSU-UNESCO; Forty Years of Cooperation*, 1986.
3 For International Biological Networks see *International Biosciences Networks*, IBN, 1993 Brochure, compiled by A N Rao, Asian Network of Biological Sciences, Singapore, 193.
4 List of UN agencies to be found in UN handbook reference 1.
5 GARP/JOC *Report of 8th Meeting of the ICSU General Committee*, Budapest, 1977, GARP Report p. 200, JOC Report p. 61.

16 Membership

1 T Younes in *Biology International*, 22, January 1991, pp. 2–9.
2 Linnean Society of London, founded 1788, oldest extant scientific society devoted to natural history. Relevant to the present work as being the repository of the archives of the IBP.
3 See IUBS General Assembly Reports for listing of IUBS members.
4 Boyd-Orr, (Lord). See chapter 10, on nutrition.
5 Berger, Hans (21 May 1873–1 June 1941). See *Dictionary of Scientific Biography*.
6 See Appendix 2: list of members of ICSU.
7 Third World Academy of Sciences. Creation proposed by Nobel Laureate Abdus Salam of Pakistan, 6 October 1981, at Pontifical Academy of Sciences. Established in 1983 as international forum for distinguished scientists from Third World. Became Scientific Associate of ICSU, 1984. Granted NGO status by UN Economic and Social Council, 1985. Administration of funds and staff taken over by UNESCO 1991. Membership at 1993 was 325.

17 Ringberg (1985) to Visegrad (1990): Self-examination

1 I Gandhi. Unesco Madras meeting on research and human needs.
2 Ringberg Conference Report, *New Agenda for International Science*, p. 60.
3 International Social Science Council ISSC; formed 1952 'for the advancement of the social sciences throughout the world'. Headquarters in Paris (1 rue Miollis, Paris 75007).
4 Thurau, K, Treasurer, ICSU 1984–90.
5 World Population Conference Report.
6 Ringberg Conference Report, p. 89, *Intellectual spectrum and reach of ICSU*.
7 Ringberg Conference Report, p. 117.
8 See chapter 15.
9 Report of Ringberg meeting, *International Science and the Role of ICSU*, p. 135.

18 The road to Rio and beyond

1 Carson, Rachel, *Silent Spring*, 1963.
2 See note 6 below.

3 El Niño, the warm current in the Pacific the variability of which has great economic and environmental effects throughout the western seaboard regions of South America.

4 *UNEP Information no 47*, Nairobi, UN, 5 June 1979. Reprinted in *Geography, Resources and Environment: Selected Writings of G F White*, eds. Kates, R W, and Burton, I, University of Chicago Press, Chicago and London, 1986, vol. I, pp. 414,416.

5 Quoted by T F Malone in: 'How it all began', Global Change Newsletter 18, IGBP, June 1994, p.8.

6 T F Malone and J G Roederer (eds.). *Global Change*, ICSU Press, 1984, p. xiv.

7 Ibidem W S Fyfe, p. 351.

8 IGBP Report no 12: *The Initial Core Projects*, 1990, pp. 1–5.

9 IGBP Report no 28: *IGBP in Action, Work Plan* 1994–1998, pp. 10 11.

10 *ICSU Year Book 1994*, p. 11.

11 Maurice Strong (born 29 April 1929), original scientific training in chemistry, long industrial and Canadian Civil Service experience.

12 IGFA Meeting *Report no 5*, 1994, p. 27.

13 *An Agenda for Science for Environment and Development into the 21st Century based on a Conference held in Vienna, Austria, in November 1991*, eds. Dooge, J C I, Goodman, G T, la Rivière, J W M, Marton-Lefèvre, J, O'Riordan, T, Praderie, F, p. 53.

14 Menon speech at UNCED: in *Science International*, 49–50, 1992, pp. 12–13.

15 *Understanding our own Planet, an Overview of Major International Scientific Activities*, ICSU Brochure, Paris, 1993, pp. 24–5.

Further reading

Baker, F W G, International Polar Year: a century of interdisciplinary co-operation, *Interdisciplinary Science Reviews*, 7(4), 1982, pp. 270–81.

Baker, F W G, Co-operation among non-governmental organizations in fostering oceanic research, *Impact*, 1983, 3(4), pp. 293–99.

Baker, F W G, *ICSU-UNESCO: forty years of cooperation*, UNESCO, 1986.

Baker, F W G, *ICSU: a Brief Survey*, ICSU, 1986 (and 1988).

Baker, F W G, First International Polar Year (1882–3) (From *Polar Record*, 21(132), September 1982).

Blaauw, Adriaan, *History of IAU*, Dordrecht/London 1994.

Brown, Harrison, (Interview with Harrison Brown), *Chemical and Engineering News*, 4 October 1976, pp. 18–19.

Bullis, H, *Political Legacy of the IGY: USA Committee on Foreign Affairs*, Library of Congress, 1973.

Crawford, Elisabeth, The universe of international science 1880–1939 in: ed. Frängsmyr, *Solomon's House Revisited*, 1990.

Fogg, G E, *A History of Antarctic Science*, Cambridge, 1992.

Frängsmyr, T (ed.), *Solomon's House Revisited*. Nobel Symposium 75. Science History Publications, USA, 1990.

Harrison, J M, The roots of IUGS, *Episodes*, Geological Newsletter 1, International Union of Geological Sciences, 1978, pp. 21–3.

Huxley, Julian, *Memories*, London, 1970.

Huxley, Julian, *Memories II*, London, 1973.

ICSTI, Review of present and past activities, ICSTI, Paris, 8 February 1993 (ref. 111/93/MO/dm).

ICSU, Ringberg Conference: International Science and the role of ICSU: a contemporary agenda, 7–9 October 1985 (1986).

IDNDR, Report of IDNDR to Santiago General Assembly of ICSU, 1993.

IGCP, International Conference of Experts for Preparing and International Geological Correlation Programme, Paris, 19–28 October 1971 (Report 1972).

Johnson, Stanley P (ed.), *United Nations: The Earth Summit. The United Nations Conference on Environment and Development (UNCED). 1992, London, 1993.*

Jones, Graham, *Role of Science and Technology in Developing Countries*, Oxford University Press for ICSU, 1971.

Keating, Michael, *The Earth Summit's Agenda for Change*, Geneva 1993.

Malone, T F, Mission to Planet Earth, *Environment*, 26(8) pp. 6–11, 39–42.

Malone, T F, and Corell, Robert, Mission to Planet Earth revisited, *Environment*, 31(3), pp. 7–12, 33–5.

Nicolet, Marcel, The International Geophysical Year (1957–1958): great achievements and minor obstacles, *Geo-Journal*, 8(4), 1984, pp. 303–20.

Palais de la Découverte, *Commemoration des Années Polaire et de l'Année Géophysique*; Palais de la Découverte, Paris, 1984.

Quarrie, Joyce (ed.), *Earth Summit 1992: the United Nations Conference on Environment and Development*, Rio, 1992. London 1992.

Rao, A N, International Biosciences Networks: Initial document giving origin and objectives, prepared by A N Rao for ICSU/UNESCO, Singapore, June 1996, pp. 40.

Ruttenberg, S, WMO Bulletin, ICSU World Data Centres, 42(2), 1993, pp. 130–6.

Schroeder-Gudehus, B, *Les scientifiques et la paix*, Montreal, 1978.

UNISIST, Intergovernmental Conference for the Establishment of a World Science Information System (Paris 4–8 October 1971), UNESCO, 1971.

URSI, For history of URSI see: *Electrical Trades Directory, Blue Book*, 1917, p. 335, International Commission for Scientific Radiotelegraphic Researches (gives a statement of the position immediately before and during the early years of WWI). *URSI Golden Jubilee Memorial*: 1963: articles by E V Appleton, R L Smith-Rose, J H Dellinger, W J G Beynon, H W S Massey. URSI 1919–79. *Proceedings of the 60th Anniversary Colloquium (17–18 September 1979)*, ed. C M Minnis, URSI, Brussels, 1979, articles by L Bossy, B Decault, J Groszkowski, G Beynon, pp. 59–88.

Williams, T I (ed.), *History of Technology*, vols. VI and VII, Oxford, 1978.

WOCE, International Origins of WOCE, *WOCE Newsletters* 1, October 1985, pp. 2–5.

Worthington, E Barton, *The Ecological Century*, Oxford, 1983.

Worthington, E Barton (ed.), *The Evolution of IBP*, Cambridge, 1973.

Index